Universitext

T0155803

For further volumes:
http://www.springer.com/series/223

Benjamin Steinberg

# Representation Theory of Finite Groups

## An Introductory Approach

Benjamin Steinberg
School of Mathematics and Statistics
Carleton University
Ottawa, ON K1S 5B6
Canada

Department of Mathematics
The City College of New York
NAC 8/133, Convent Avenue at 138th Street
New York, NY 10031
USA
bsteinbg@math.carleton.ca

*Editorial board*:
Sheldon Axler, San Francisco State University, San Francisco, CA, USA
Vincenzo Capasso, University of Milan, Milan, Italy
Carles Casacuberta, Universitat de Barcelona, Barcelona, Spain
Angus MacIntyre, Queen Mary, University of London, London, UK
Kenneth Ribet, University of California, Berkeley, CA, USA
Claude Sabbah, Ecole Polytechnique, Palaiseau, France
Endre Süli, Oxford University, Oxford, UK
Wojbor Woyczynski, Case Western Reserve University, Cleveland, OH, USA

ISSN 0172-5939 e-ISSN 2191-6675
ISBN 978-1-4614-0775-1 e-ISBN 978-1-4614-0776-8
DOI 10.1007/978-1-4614-0776-8
Springer New York Dordrecht Heidelberg London

Library of Congress Control Number: 2011938143

© Springer Science+Business Media, LLC 2012
All rights reserved. This work may not be translated or copied in whole or in part without the written permission of the publisher (Springer Science+Business Media, LLC, 233 Spring Street, New York, NY 10013, USA), except for brief excerpts in connection with reviews or scholarly analysis. Use in connection with any form of information storage and retrieval, electronic adaptation, computer software, or by similar or dissimilar methodology now known or hereafter developed is forbidden.
The use in this publication of trade names, trademarks, service marks, and similar terms, even if they are not identified as such, is not to be taken as an expression of opinion as to whether or not they are subject to proprietary rights.

Printed on acid-free paper

Springer is part of Springer Science+Business Media (www.springer.com)

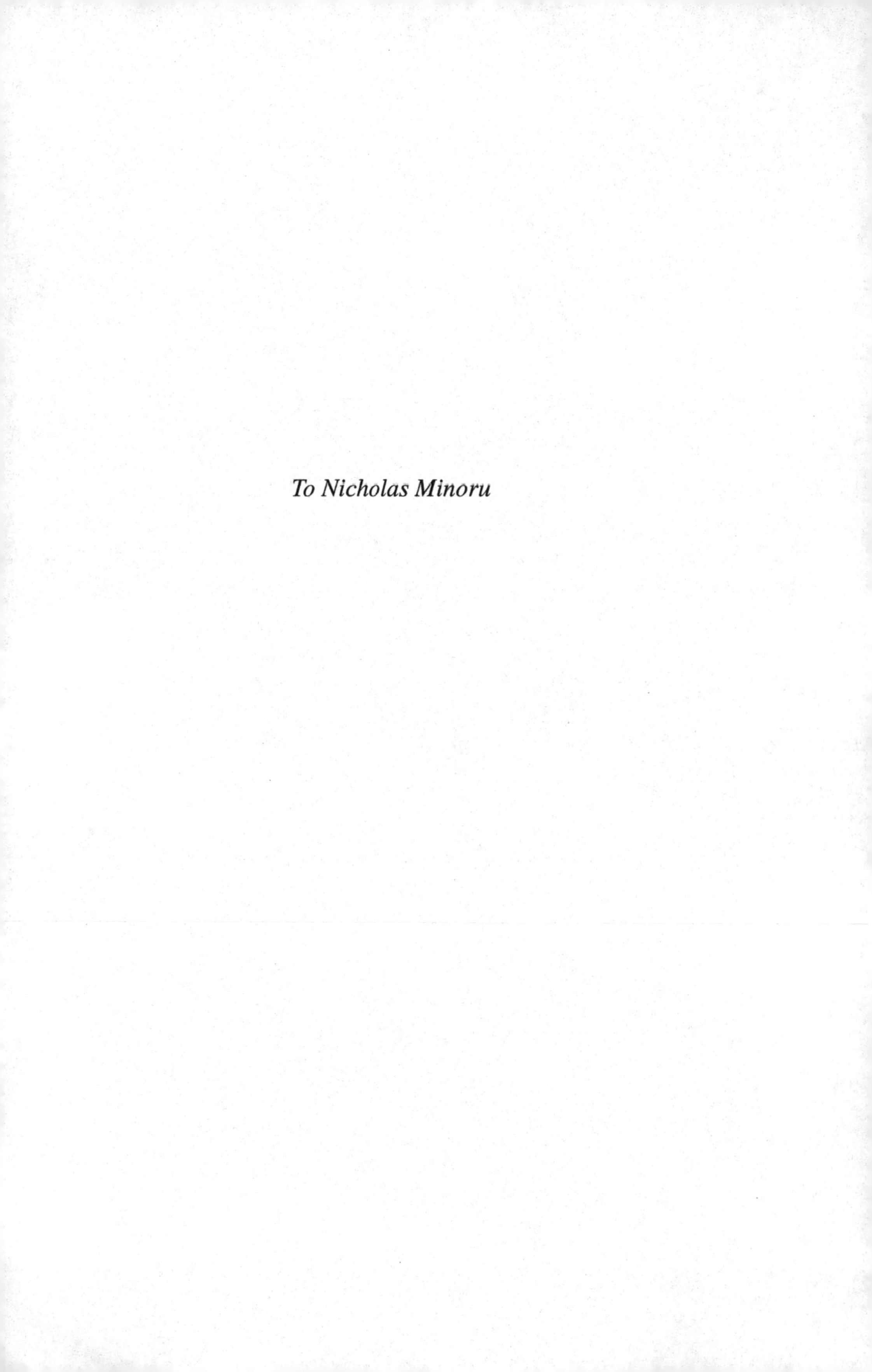

*To Nicholas Minoru*

# Preface

A student's first encounter with abstract algebra is usually in the form of linear algebra. This is typically followed by a course in group theory. Therefore, there is no reason a priori that an undergraduate could not learn group representation theory before taking a more advanced algebra course that covers module theory. In fact, group representation theory arguably could serve as a strong motivation for studying module theory over non-commutative rings. Also group representation theory has applications in diverse areas such as physics, statistics, and engineering, to name a few. One cannot always expect students from these varied disciplines to have studied module theory at the level needed for the "modern" approach to representation theory via the theory of semisimple algebras. Nonetheless, it is difficult, if not impossible, to find an undergraduate text on representation theory assuming little beyond linear algebra and group theory in prerequisites, and also assuming only a modest level of mathematical maturity.

During the Winter term of 2008, I taught a fourth year undergraduate/first year graduate course at Carleton University, which also included some third year students (and even one exceptionally talented second year student). After a bit of searching, I failed to find a text that matched the background and level of mathematical sophistication of my students. Faced with this situation, I decided to provide instead my own course notes, which have since evolved into this text. The goal was to present a gentle and leisurely introduction to group representation theory, at a level that would be accessible to students who have not yet studied module theory and who are unfamiliar with the more sophisticated aspects of linear algebra, such as tensor products. For this reason, I chose to avoid completely the Wedderburn theory of semisimple algebras. Instead, I have opted for a Fourier analytic approach. This sort of approach is normally taken in books with a more analytic flavor; such books, however, invariably contain material on the representation theory of compact groups, something else that I would consider beyond the scope of an undergraduate text. So here I have done my best to blend the analytic and the algebraic viewpoints in order to keep things accessible. For example, Frobenius reciprocity is treated from a character point of view to avoid use of the tensor product.

The only background required for most of this book is a basic knowledge of linear algebra and group theory, as well as familiarity with the definition of a ring. In particular, we assume familiarity with the symmetric group and cycle notation. The proof of Burnside's theorem makes use of a small amount of Galois theory (up to the fundamental theorem) and so should be skipped if used in a course for which Galois theory is not a prerequisite. Many things are proved in more detail than one would normally expect in a textbook; this was done to make things easier on undergraduates trying to learn what is usually considered graduate level material.

The main topics covered in this book include: character theory; the group algebra and Fourier analysis; Burnside's $pq$-theorem and the dimension theorem; permutation representations; induced representations and Mackey's theorem; and the representation theory of the symmetric group. The book ends with a chapter on applications to probability theory via random walks on groups.

It should be possible to present this material in a one semester course. Chapters 2–5 should be read by everybody; it covers the basic character theory of finite groups. The first two sections of Chap. 6 are also recommended for all readers; the reader who is less comfortable with Galois theory can then skip the last section of this chapter and move on to Chap. 7 on permutation representations, which is needed for Chaps. 8–10. Chapter 10, on the representation theory of the symmetric group, can be read immediately after Chap. 7. The final chapter, Chap. 11, provides an introduction to random walks on finite groups. It is intended to serve as a non-trivial application of representation theory, rather than as part of the core material of the book, and should therefore be taken as optional for those interested in the purely algebraic aspects of the theory. Chapter 11 can be read directly after Chap. 5, as it relies principally on Fourier analysis on abelian groups.

Although this book is envisioned as a text for an advanced undergraduate or introductory graduate level course, it is also intended to be of use for physicists, statisticians, and mathematicians who may not be algebraists, but need group representation theory for their work.

While preparing this book I have relied on a number of classical references on representation theory, including [5–7, 10, 15, 20, 21]. For the representation theory of the symmetric group I have drawn from [7, 12, 13, 16, 17, 19]; the approach is due to James [17]. Good references for applications of representation theory to computing eigenvalues of graphs and random walks are [3, 6, 7]. Chapter 11, in particular, owes much of its presentation to [7] and [3]. Discrete Fourier analysis and its applications can be found in [3, 7, 22].

Thanks are due to the following people for their input and suggestions: Ariane Masuda, Paul Mezo, and Martin Steinberg.

Ottawa Benjamin Steinberg

# Contents

# List of Figures

# List of Tables

# Chapter 1
# Introduction

The representation theory of finite groups is a subject going back to the late eighteen hundreds. The pioneers in the subject were G. Frobenius, I. Schur, and W. Burnside. Modern approaches tend to make heavy use of module theory and the Wedderburn theory of semisimple algebras. But the original approach, which nowadays can be thought of as via discrete Fourier analysis, is much more easily accessible and can be presented, for instance, in an undergraduate course. The aim of this text is to exposit the essential ingredients of the representation theory of finite groups over the complex numbers assuming only knowledge of linear algebra and undergraduate group theory, and perhaps a minimal familiarity with ring theory.

The original purpose of representation theory was to serve as a powerful tool for obtaining information about finite groups via the methods of linear algebra, e.g., eigenvalues, inner product spaces, and diagonalization. The first major triumph of representation theory was Burnside's $pq$-theorem. This theorem states that a non-abelian group of order $p^a q^b$ with $p, q$ prime cannot be simple, or equivalently, that every finite group of order $p^a q^b$ with $p, q$ prime is solvable. It was not until much later [2, 14] that purely group theoretic proofs were found. Representation theory went on to play an indispensable role in the classification of finite simple groups.

However, representation theory is much more than just a means to study the structure of finite groups. It is also a fundamental tool with applications to many areas of mathematics and statistics, both pure and applied. For instance, sound compression is very much based on the fast Fourier transform for finite abelian groups. Fourier analysis on finite groups also plays an important role in probability and statistics, especially in the study of random walks on groups, such as card shuffling and diffusion processes [3, 7], and in the analysis of data [7, 8]; random walks are considered in the last chapter of the book. Applications of representation theory to graph theory, and in particular to the construction of expander graphs, can be found in [6]. Some applications along these lines, especially toward the computation of eigenvalues of Cayley graphs, are given in this text.

B. Steinberg, *Representation Theory of Finite Groups: An Introductory Approach*, Universitext, DOI 10.1007/978-1-4614-0776-8_1,
© Springer Science+Business Media, LLC 2012

# Chapter 2
# Review of Linear Algebra

This chapter reviews the linear algebra that we shall assume throughout the book. Proofs of standard results are mostly omitted. The reader can consult a linear algebra text such as [4] for details. In this book all vector spaces considered will be finite dimensional over the field $\mathbb{C}$ of complex numbers.

## 2.1 Basic Definitions and Notation

This section introduces some basic notions from linear algebra. We start with some notation, not all of which belongs to linear algebra. Let $V$ and $W$ be vector spaces.

- If $X$ is a set of vectors, then $\mathbb{C}X = \operatorname{Span} X$.
- $M_{mn}(\mathbb{C}) = \{m \times n \text{ matrices with entries in } \mathbb{C}\}$.
- $M_n(\mathbb{C}) = M_{nn}(\mathbb{C})$.
- $\operatorname{Hom}(V, W) = \{A \colon V \longrightarrow W \mid A \text{ is a linear map}\}$.
- $\operatorname{End}(V) = \operatorname{Hom}(V, V)$ (the *endomorphism ring* of $V$).
- $GL(V) = \{A \in \operatorname{End}(V) \mid A \text{ is invertible}\}$ (known as the *general linear group of* $V$).
- $GL_n(\mathbb{C}) = \{A \in M_n(\mathbb{C}) \mid A \text{ is invertible}\}$.
- The identity matrix/linear transformation is denoted $I$, or $I_n$ if we wish to emphasize the dimension $n$.
- $\mathbb{Z}$ is the ring of integers.
- $\mathbb{N}$ is the set of non-negative integers.
- $\mathbb{Q}$ is the field of rational numbers.
- $\mathbb{R}$ is the field of real numbers.
- $\mathbb{Z}/n\mathbb{Z} = \{[0], \ldots, [n-1]\}$ is the ring of integers modulo $n$.
- $R^*$ denotes the group of units (i.e., invertible elements) of a ring $R$.
- $S_n$ is the group of permutations of $\{1, \ldots, n\}$, i.e., the *symmetric group* on $n$ letters.
- The identity permutation is denoted $Id$.

B. Steinberg, *Representation Theory of Finite Groups: An Introductory Approach*, Universitext, DOI 10.1007/978-1-4614-0776-8_2,
© Springer Science+Business Media, LLC 2012

Elements of $\mathbb{C}^n$ will be written as $n$-tuples or as column vectors, as is convenient.

If $A \in M_{mn}(\mathbb{C})$, we sometimes write $A_{ij}$ for the entry in row $i$ and column $j$. We may also write $A = (a_{ij})$ to mean the matrix with $a_{ij}$ in row $i$ and column $j$. If $k, \ell, m$, and $n$ are natural numbers, then matrices in $M_{mk,\ell n}(\mathbb{C})$ can be viewed as $m \times n$ block matrices with blocks in $M_{k\ell}(\mathbb{C})$. If we view an $mk \times \ell n$ matrix $A$ as a block matrix, then we write $[A]_{ij}$ for the $k \times \ell$ matrix in the $i, j$ block, for $1 \leq i \leq m$ and $1 \leq j \leq n$.

**Definition 2.1.1 (Coordinate vector).** If $V$ is a vector space with basis $B = \{b_1, \ldots, b_n\}$ and $v = c_1b_1 + \cdots + c_nb_n$ is a vector in $V$, then the *coordinate vector* of $v$ with respect to the basis $B$ is the vector $[v]_B = (c_1, \ldots, c_n) \in \mathbb{C}^n$. The map $T\colon V \longrightarrow \mathbb{C}^n$ given by $Tv = [v]_B$ is a vector space isomorphism that we sometimes call *taking coordinates* with respect to $B$.

Suppose that $T\colon V \longrightarrow W$ is a linear transformation and $B, B'$ are bases for $V, W$, respectively. Let $B = \{v_1, \ldots, v_n\}$ and $B' = \{w_1, \ldots, w_m\}$. Then the *matrix of* $T$ with respect to the bases $B, B'$ is the $m \times n$ matrix $[T]_{B,B'}$ whose $j$th column is $[Tv_j]_{B'}$. In other words, if

$$Tv_j = \sum_{i=1}^{m} a_{ij} w_i,$$

then $[T]_{B,B'} = (a_{ij})$. When $V = W$ and $B = B'$, then we write simply $[T]_B$ for $[T]_{B,B}$.

The *standard basis* for $\mathbb{C}^n$ is the set $\{e_1, \ldots, e_n\}$ where $e_i$ is the vector with 1 in the $i$th coordinate and 0 in all other coordinates. So when $n = 3$, we have

$$e_1 = (1,0,0), \quad e_2 = (0,1,0), \quad e_3 = (0,0,1).$$

Throughout we will abuse the distinction between $\mathrm{End}(\mathbb{C}^n)$ and $M_n(\mathbb{C})$ and the distinction between $GL(\mathbb{C}^n)$ and $GL_n(\mathbb{C})$ by identifying a linear transformation with its matrix with respect to the standard basis.

Suppose $\dim V = n$ and $\dim W = m$. Then by choosing bases for $V$ and $W$ and sending a linear transformation to its matrix with respect to these bases we see that:

$$\begin{aligned} \mathrm{End}(V) &\cong M_n(\mathbb{C}); \\ GL(V) &\cong GL_n(\mathbb{C}); \\ \mathrm{Hom}(V, W) &\cong M_{mn}(\mathbb{C}). \end{aligned}$$

Notice that $GL_1(\mathbb{C}) \cong \mathbb{C}^*$ and so we shall always work with the latter. We indicate $W$ is a subspace of $V$ by writing $W \leq V$.

If $W_1, W_2 \leq V$, then by definition

$$W_1 + W_2 = \{w_1 + w_2 \mid w_1 \in W_1, w_2 \in W_2\}.$$

This is the smallest subspace of $V$ containing $W_1$ and $W_2$. If, in addition, $W_1 \cap W_2 = \{0\}$, then $W_1 + W_2$ is called a *direct sum*, written $W_1 \oplus W_2$. As vector spaces, $W_1 \oplus W_2 \cong W_1 \times W_2$ via the map $W_1 \times W_2 \longrightarrow W_1 \oplus W_2$ given by $(w_1, w_2) \mapsto w_1 + w_2$. In fact, if $V$ and $W$ are any two vector spaces, one can form their *external direct sum* by setting $V \oplus W = V \times W$. Note that

$$\dim(W_1 \oplus W_2) = \dim W_1 + \dim W_2.$$

More precisely, if $B_1$ is a basis for $W_1$ and $B_2$ is a basis for $W_2$, then $B_1 \cup B_2$ is a basis for $W_1 \oplus W_2$.

## 2.2 Complex Inner Product Spaces

Recall that if $z = a + bi \in \mathbb{C}$, then its *complex conjugate* is $\overline{z} = a - bi$. In particular, $z\overline{z} = a^2 + b^2 = |z|^2$. An *inner product* on $V$ is a map

$$\langle \cdot, \cdot \rangle \colon V \times V \longrightarrow \mathbb{C}$$

such that, for $v, w, v_1, v_2 \in V$ and $c_1, c_2 \in \mathbb{C}$:

- $\langle c_1 v_1 + c_2 v_2, w \rangle = c_1 \langle v_1, w \rangle + c_2 \langle v_2, w \rangle$;
- $\langle w, v \rangle = \overline{\langle v, w \rangle}$;
- $\langle v, v \rangle \geq 0$ and $\langle v, v \rangle = 0$ if and only if $v = 0$.

A vector space equipped with an inner product is called an *inner product space*. The *norm* $\|v\|$ of a vector $v$ in an inner product space is defined by $\|v\| = \sqrt{\langle v, v \rangle}$.

*Example 2.2.1.* The *standard inner product* on $\mathbb{C}^n$ is given by

$$\langle (a_1, \ldots, a_n), (b_1, \ldots, b_n) \rangle = \sum_{i=1}^{n} a_i \overline{b_i}.$$

Two important properties of inner products are the *Cauchy–Schwarz inequality*

$$|\langle v, w \rangle| \leq \|v\| \cdot \|w\|$$

and the *triangle inequality*

$$\|v + w\| \leq \|v\| + \|w\|.$$

Recall that two vectors $v, w$ in an inner product space $V$ are said to be *orthogonal* if $\langle v, w \rangle = 0$. A subset of $V$ is called *orthogonal* if its elements are pairwise orthogonal. If, in addition, the norm of each vector is 1, the set is termed *orthonormal*. An orthogonal set of non-zero vectors is linearly independent. In particular, any orthonormal set is linearly independent.

Every inner product space has an orthonormal basis. One can obtain an orthonormal basis from an arbitrary basis using the Gram–Schmidt process [4, Theorem 15.9]. If $B = \{e_1, \ldots, e_n\}$ is an orthonormal basis for an inner product space $V$ and $v \in V$, then

$$v = \langle v, e_1\rangle e_1 + \cdots + \langle v, e_n\rangle e_n$$

In other words,

$$[v]_B = (\langle v, e_1\rangle, \ldots, \langle v, e_n\rangle).$$

*Example 2.2.2.* For a finite set $X$, the set $\mathbb{C}^X = \{f : X \longrightarrow \mathbb{C}\}$ is a vector space with pointwise operations. Namely, one defines

$$(f+g)(x) = f(x) + g(x);$$
$$(cf)(x) = cf(x).$$

For each $x \in X$, define a function $\delta_x : X \longrightarrow \mathbb{C}$ by

$$\delta_x(y) = \begin{cases} 1 & x = y \\ 0 & x \neq y. \end{cases}$$

There is a natural inner product on $\mathbb{C}^X$ given by

$$\langle f, g\rangle = \sum_{x \in X} f(x)\overline{g(x)}.$$

The set $\{\delta_x \mid x \in X\}$ is an orthonormal basis with respect to this inner product. If $f \in \mathbb{C}^X$, then its unique expression as a linear combination of the $\delta_x$ is given by

$$f = \sum_{x \in X} f(x)\delta_x.$$

Consequently, $\dim \mathbb{C}^X = |X|$.

Direct sum decompositions are easy to obtain in inner product spaces. If $W \leq V$, then the *orthogonal complement* of $W$ is the subspace

$$W^\perp = \{v \in V \mid \langle v, w\rangle = 0 \text{ for all } w \in W\}.$$

**Proposition 2.2.3.** *Let $V$ be an inner product space and $W \leq V$. Then there is a direct sum decomposition $V = W \oplus W^\perp$.*

*Proof.* First, if $w \in W \cap W^\perp$ then $\langle w, w\rangle = 0$ implies $w = 0$; so $W \cap W^\perp = \{0\}$. Let $v \in V$ and suppose that $\{e_1, \ldots, e_m\}$ is an orthonormal basis for $W$. Put $\hat{v} = \langle v, e_1\rangle e_1 + \cdots + \langle v, e_m\rangle e_m$ and $z = v - \hat{v}$. Then $\hat{v} \in W$. We claim that $z \in W^\perp$.

To prove this, it suffices to show $\langle z, e_i \rangle = 0$ for all $i = 1, \ldots, m$. To this effect we compute

$$\langle z, e_i \rangle = \langle v, e_i \rangle - \langle \hat{v}, e_i \rangle = \langle v, e_i \rangle - \langle v, e_i \rangle = 0$$

because $\{e_1, \ldots, e_m\}$ is an orthonormal set. As $v = \hat{v} + z$, it follows that $V = W + W^{\perp}$. This completes the proof. $\square$

We continue to assume that $V$ is an inner product space.

**Definition 2.2.4 (Unitary operator).** A linear operator $U \in GL(V)$ is said to be *unitary* if

$$\langle Uv, Uw \rangle = \langle v, w \rangle$$

for all $v, w \in V$.

Notice that if $U$ is unitary and $v \in \ker U$, then $0 = \langle Uv, Uv \rangle = \langle v, v \rangle$ and so $v = 0$. Thus unitary operators are invertible. The set $U(V)$ of unitary maps is a subgroup of $GL(V)$.

If $A = (a_{ij}) \in M_{mn}(\mathbb{C})$ is a matrix, then its *transpose* is the matrix $A^T = (a_{ji}) \in M_{nm}(\mathbb{C})$. The *conjugate* of $A$ is $\overline{A} = (\overline{a_{ij}})$. The *conjugate-transpose* or *adjoint* of $A$ is the matrix $A^* = \overline{A^T}$. One can verify directly the equality $(AB)^* = B^*A^*$. Routine computation shows that if $v \in \mathbb{C}^n$ and $w \in \mathbb{C}^m$, then

$$\langle Av, w \rangle = \langle v, A^*w \rangle \tag{2.1}$$

where we use the standard inner product on $\mathbb{C}^m$ and $\mathbb{C}^n$. Indeed, viewing vectors as column vectors one has $\langle v_1, v_2 \rangle = \overline{v_1^* v_2}$ and so $\langle Av, w \rangle = \overline{(Av)^* w} = \overline{v^*(A^*w)} = \langle v, A^*w \rangle$.

With respect to the standard inner product on $\mathbb{C}^n$, the linear transformation associated to a matrix $A \in GL_n(\mathbb{C})$ is unitary if and only if $A^{-1} = A^*$ [4, Theorem 32.7]; such a matrix is thus called *unitary*. We denote by $U_n(\mathbb{C})$ the group of all $n \times n$ unitary matrices. A matrix $A \in M_n(\mathbb{C})$ is called *self-adjoint* if $A^* = A$. A matrix $A$ is *symmetric* if $A^T = A$. If $A$ has real entries, then $A$ is self-adjoint if and only if $A$ is symmetric.

More generally, if $T$ is a linear operator on an inner product space $V$, then $T^* \colon V \longrightarrow V$ is the unique linear operator satisfying $\langle Tv, w \rangle = \langle v, T^*w \rangle$ for all $v, w \in V$. It is called the *adjoint* of $T$. If $B$ is an orthonormal basis for $V$, then $[T^*]_B = [T]_B^*$. The operator $T$ is *self-adjoint* if $T = T^*$, or equivalently if the matrix of $T$ with respect to some (equals any) orthonormal basis of $V$ is self-adjoint.

## 2.3 Further Notions from Linear Algebra

If $X \subseteq \mathrm{End}(V)$ and $W \leq V$, then $W$ is called $X$*-invariant* if, for any $A \in X$ and any $w \in W$, one has $Aw \in W$, i.e., $XW \subseteq W$.

A key example comes from the theory of eigenvalues and eigenvectors. Recall that $\lambda \in \mathbb{C}$ is an *eigenvalue* of $A \in \mathrm{End}(V)$ if $\lambda I - A$ is not invertible, or in other words, if $Av = \lambda v$ for some $v \neq 0$. The *eigenspace* corresponding to $\lambda$ is the set

$$V_\lambda = \{v \in V \mid Av = \lambda v\},$$

which is a subspace of $V$. Note that if $v \in V_\lambda$, then $A(Av) = A(\lambda v) = \lambda Av$, so $Av \in V_\lambda$. Thus $V_\lambda$ is $A$-invariant. On the other hand, if $W \leq V$ is $A$-invariant with $\dim W = 1$ (that is, $W$ is a line), then $W \subseteq V_\lambda$ for some $\lambda$. In fact, if $w \in W \setminus \{0\}$, then $\{w\}$ is a basis for $W$. Since $Aw \in W$, we have that $Aw = \lambda w$ for some $\lambda \in \mathbb{C}$. So $w$ is an eigenvector with eigenvalue $\lambda$, whence $w \in V_\lambda$. Thus $W \subseteq V_\lambda$.

The *trace* of a matrix $A = (a_{ij})$ is defined by

$$\mathrm{Tr}(A) = \sum_{i=1}^{n} a_{ii}.$$

Some basic facts concerning the trace function $\mathrm{Tr}\colon M_n(\mathbb{C}) \longrightarrow \mathbb{C}$ are that Tr is linear and $\mathrm{Tr}(AB) = \mathrm{Tr}(BA)$. Consequently, $\mathrm{Tr}(PAP^{-1}) = \mathrm{Tr}(P^{-1}PA) = \mathrm{Tr}(A)$ for any invertible matrix $P$. In particular, if $T \in \mathrm{End}(V)$, then $\mathrm{Tr}(T)$ makes sense: choose any basis for the vector space $V$ and compute the trace of the associated matrix.

The *determinant* $\det A$ of a matrix is defined as follows:

$$\det A = \sum_{\sigma \in S_n} \mathrm{sgn}(\sigma) \cdot a_{1\sigma(1)} \cdots a_{n\sigma(n)}.$$

We recall that

$$\mathrm{sgn}(\sigma) = \begin{cases} 1 & \sigma \text{ is even} \\ -1 & \sigma \text{ is odd.} \end{cases}$$

The key properties of the determinant that we shall use are:

- $\det A \neq 0$ if and only if $A \in GL_n(\mathbb{C})$;
- $\det(AB) = \det A \cdot \det B$;
- $\det(A^{-1}) = (\det A)^{-1}$.

In particular, one has $\det(PAP^{-1}) = \det A$ and so we can define, for any $T \in \mathrm{End}(V)$, the determinant by choosing a basis for $V$ and computing the determinant of the corresponding matrix for $T$.

The *characteristic polynomial* $p_A(x)$ of a linear operator $A$ on an $n$-dimensional vector space $V$ is given by $p_A(x) = \det(xI - A)$. This is a monic polynomial of degree $n$ (i.e., has leading coefficient 1) and the roots of $p_A(x)$ are precisely the eigenvalues of $A$. The classical Cayley–Hamilton theorem says that any linear operator is a zero of its characteristic polynomial [4, Corollary 24.7].

**Theorem 2.3.1 (Cayley–Hamilton).** *Let $p_A(x)$ be the characteristic polynomial of $A$. Then $p_A(A) = 0$.*

If $A \in \mathrm{End}(V)$, *the minimal polynomial* of $A$, denoted $m_A(x)$, is the smallest degree monic polynomial $f(x)$ such that $f(A) = 0$.

**Proposition 2.3.2.** *If $q(A) = 0$ then $m_A(x) \mid q(x)$.*

*Proof.* Write $q(x) = m_A(x)f(x) + r(x)$ with either $r(x) = 0$, or $\deg(r(x)) < \deg(m_A(x))$. Then

$$0 = q(A) = m_A(A)f(A) + r(A) = r(A).$$

By minimality of $m_A(x)$, we conclude that $r(x) = 0$. □

**Corollary 2.3.3.** *If $p_A(x)$ is the characteristic polynomial of $A$, then $m_A(x)$ divides $p_A(x)$.*

The relevance of the minimal polynomial is that it provides a criterion for diagonalizability of a matrix, among other things. A standard result from linear algebra is the following characterization of diagonalizable matrices, cf. [4, Theorem 23.11].

**Theorem 2.3.4.** *A matrix $A \in M_n(\mathbb{C})$ is diagonalizable if and only if $m_A(x)$ has no repeated roots.*

*Example 2.3.5.* The diagonal matrix

$$A = \begin{bmatrix} 3 & 0 & 0 \\ 0 & 1 & 0 \\ 0 & 0 & 1 \end{bmatrix}$$

has $m_A(x) = (x-1)(x-3)$, whereas $p_A(x) = (x-1)^2(x-3)$. On the other hand, the matrix

$$B = \begin{bmatrix} 1 & 1 \\ 0 & 1 \end{bmatrix}$$

has $m_B(x) = (x-1)^2 = p_B(x)$ and so is not diagonalizable.

One of the main results from linear algebra is the spectral theorem for self-adjoint matrices. We sketch a proof since it is indicative of several proofs later in the text.

**Theorem 2.3.6 (Spectral Theorem).** *Let $A \in M_n(\mathbb{C})$ be self-adjoint. Then there is a unitary matrix $U \in U_n(\mathbb{C})$ such that $U^*AU$ is diagonal. Moreover, the eigenvalues of $A$ are real.*

*Proof.* First we verify that the eigenvalues are real. Let $\lambda$ be an eigenvalue of $A$ with corresponding eigenvector $v$. Then

$$\lambda\langle v, v\rangle = \langle Av, v\rangle = \langle v, A^*v\rangle = \langle v, Av\rangle = \overline{\langle Av, v\rangle} = \overline{\lambda}\langle v, v\rangle$$

and hence $\lambda = \overline{\lambda}$ because $\langle v, v\rangle > 0$. Thus $\lambda$ is real.

To prove the remainder of the theorem, it suffices to show that $\mathbb{C}^n$ has an orthonormal basis $B$ of eigenvectors for $A$. One can then take $U$ to be a matrix whose columns are the elements of $B$. We proceed by induction on $n$, the case $n = 1$ being trivial. Assume the theorem is true for all dimensions smaller than $n$. Let $\lambda$ be an eigenvalue of $A$ with corresponding eigenspace $V_\lambda$. If $V_\lambda = \mathbb{C}^n$, then $A$ already is diagonal and there is nothing to prove. So we may assume that $V_\lambda$ is a proper subspace; it is of course non-zero. Then $\mathbb{C}^n = V_\lambda \oplus V_\lambda^\perp$ by Proposition 2.2.3. We claim that $V_\lambda^\perp$ is $A$-invariant. Indeed, if $v \in V_\lambda$ and $w \in V_\lambda^\perp$, then

$$\langle Aw, v\rangle = \langle w, A^*v\rangle = \langle w, Av\rangle = \langle w, \lambda v\rangle = 0$$

and so $Aw \in V_\lambda^\perp$. Note that $V_\lambda^\perp$ is an inner product space in its own right by restricting the inner product on $V$, and moreover the restriction of $A$ to $V_\lambda^\perp$ is still self-adjoint. Since $\dim V_\lambda^\perp < n$, an application of the induction hypothesis yields that $V_\lambda^\perp$ has an orthonormal basis $B'$ of eigenvectors for $A$. Let $B$ be any orthonormal basis for $V_\lambda$. Then $B \cup B'$ is an orthonormal basis for $\mathbb{C}^n$ consisting of eigenvectors for $A$, as required. □

## Exercises

**Exercise 2.1.** Suppose that $A, B \in M_n(\mathbb{C})$ are commuting matrices, i.e., $AB = BA$. Let $V_\lambda$ be an eigenspace of $A$. Show that $V_\lambda$ is $B$-invariant.

**Exercise 2.2.** Let $V$ be an $n$-dimensional vector space and $B$ a basis. Prove that the map $F\colon \operatorname{End}(V) \longrightarrow M_n(\mathbb{C})$ given by $F(T) = [T]_B$ is an isomorphism of unital rings.

**Exercise 2.3.** Let $V$ be an inner product space and let $W \leq V$ be a subspace. Let $v \in V$ and define $\hat{v} \in W$ as in the proof of Proposition 2.2.3. Prove that if $w \in W$ with $w \neq \hat{v}$, then $\|v - \hat{v}\| < \|v - w\|$. Deduce that $\hat{v}$ is independent of the choice of orthonormal basis for $W$. It is called the *orthogonal projection* of $v$ onto $W$.

**Exercise 2.4.** Prove that $(AB)^* = B^*A^*$.

**Exercise 2.5.** Prove that $\operatorname{Tr}(AB) = \operatorname{Tr}(BA)$.

**Exercise 2.6.** Let $V$ be an inner product space and let $T\colon V \longrightarrow V$ be a linear transformation. Show that $T$ is self-adjoint if and only if $V$ has an orthonormal basis of eigenvectors of $T$ and the eigenvalues of $T$ are real. (Hint: one direction is a consequence of the spectral theorem.)

**Exercise 2.7.** Let $V$ be an inner product space. Show that $U \in \operatorname{End}(V)$ is unitary if and only if $\|Uv\| = \|v\|$ for all vectors $v \in V$. (Hint: use the polarization formula $\langle v, w\rangle = 1/4\left[\|v + w\|^2 - \|v - w\|^2\right]$.)

**Exercise 2.8.** Prove that if $A \in M_n(\mathbb{C})$, then there is an upper triangular matrix $T$ and an invertible matrix $P$ such that $P^{-1}AP = T$. (Hint: use induction on $n$. Look at the proof of the spectral theorem for inspiration.)

**Exercise 2.9.** This exercise sketches a proof of the Cayley–Hamilton Theorem using a little bit of analysis.

1. Use Exercise 2.8 to reduce to the case when $A$ is an upper triangular matrix.
2. Prove the Cayley–Hamilton theorem for diagonalizable operators.
3. Identifying $M_n(\mathbb{C})$ with $\mathbb{C}^{n^2}$, show that the mapping $M_n(\mathbb{C}) \longrightarrow M_n(\mathbb{C})$ given by $A \mapsto p_A(A)$ is continuous. (Hint: the coefficients of $p_A(x)$ are polynomials in the entries of $A$.)
4. Prove that every upper triangular matrix is a limit of matrices with distinct eigenvalues (and hence diagonalizable).
5. Deduce the Cayley–Hamilton theorem.

# Chapter 3
# Group Representations

The goal of group representation theory is to study groups via their actions on vector spaces. Consideration of groups acting on sets leads to such important results as the Sylow theorems. By studying actions on vector spaces even more detailed information about a group can be obtained. This is the subject of representation theory. Our study of matrix representations of groups will lead us naturally to Fourier analysis and the study of complex-valued functions on a group. This in turn has applications to various disciplines like engineering, graph theory, and probability, just to name a few.

## 3.1 Basic Definitions and First Examples

The reader should recall from group theory that an *action* of a group $G$ on a set $X$ is by definition a homomorphism $\varphi\colon G \longrightarrow S_X$, where $S_X$ is the symmetric group on $X$. This motivates the following definition.

**Definition 3.1.1 (Representation).** A *representation* of a group $G$ is a homomorphism $\varphi\colon G \longrightarrow GL(V)$ for some (finite-dimensional) vector space $V$. The dimension of $V$ is called the *degree of* $\varphi$. We usually write $\varphi_g$ for $\varphi(g)$ and $\varphi_g(v)$, or simply $\varphi_g v$, for the action of $\varphi_g$ on $v \in V$.

*Remark 3.1.2.* We shall rarely have occasion to consider degree zero representations and so the reader can safely ignore them. That is, we shall tacitly assume in this text that representations are non-zero, although this is not formally part of the definition.

A particularly simple example of a representation is the trivial representation.

*Example 3.1.3 (Trivial representation).* The trivial representation of a group $G$ is the homomorphism $\varphi\colon G \longrightarrow \mathbb{C}^*$ given by $\varphi(g) = 1$ for all $g \in G$.

B. Steinberg, *Representation Theory of Finite Groups: An Introductory Approach*, Universitext, DOI 10.1007/978-1-4614-0776-8_3,
© Springer Science+Business Media, LLC 2012

Let us consider some other examples of degree one representations.

*Example 3.1.4.* $\varphi\colon \mathbb{Z}/2\mathbb{Z} \longrightarrow \mathbb{C}^*$ given by $\varphi([m]) = (-1)^m$ is a representation.

*Example 3.1.5.* $\varphi\colon \mathbb{Z}/4\mathbb{Z} \longrightarrow \mathbb{C}^*$ given by $\varphi([m]) = i^m$ is a representation.

*Example 3.1.6.* More generally, $\varphi\colon \mathbb{Z}/n\mathbb{Z} \longrightarrow \mathbb{C}^*$ defined by $\varphi([m]) = e^{2\pi i m/n}$ is a representation.

Let $\varphi\colon G \longrightarrow GL(V)$ be a representation of degree $n$. To a basis $B$ for $V$, we can associate a vector space isomorphism $T\colon V \longrightarrow \mathbb{C}^n$ by taking coordinates. We can then define a representation $\psi\colon G \longrightarrow GL_n(\mathbb{C})$ by setting $\psi_g = T\varphi_g T^{-1}$ for $g \in G$. If $B'$ is another basis, we have another isomorphism $S\colon V \longrightarrow \mathbb{C}^n$, and hence a representation $\psi'\colon G \longrightarrow GL_n(\mathbb{C})$ given by $\psi'_g = S\varphi_g S^{-1}$. The representations $\psi$ and $\psi'$ are related via the formula $\psi'_g = ST^{-1}\psi_g TS^{-1} = (ST^{-1})\psi_g(ST^{-1})^{-1}$. We want to think of $\varphi$, $\psi$, and $\psi'$ as all being the same representation. This leads us to the important notion of equivalence.

**Definition 3.1.7 (Equivalence).** Two representations $\varphi\colon G \longrightarrow GL(V)$ and $\psi\colon G \longrightarrow GL(W)$ are said to be *equivalent* if there exists an isomorphism $T\colon V \longrightarrow W$ such that $\psi_g = T\varphi_g T^{-1}$ for all $g \in G$, i.e., $\psi_g T = T\varphi_g$ for all $g \in G$. In this case, we write $\varphi \sim \psi$. In pictures, we have that the diagram

$$\begin{array}{ccc} V & \xrightarrow{\varphi_g} & V \\ {\scriptstyle T}\downarrow & & \downarrow{\scriptstyle T} \\ W & \xrightarrow[\psi_g]{} & W \end{array}$$

commutes, meaning that either of the two ways of going from the upper left to the lower right corner of the diagram give the same answer.

*Example 3.1.8.* Define $\varphi\colon \mathbb{Z}/n\mathbb{Z} \longrightarrow GL_2(\mathbb{C})$ by

$$\varphi_{[m]} = \begin{bmatrix} \cos\left(\dfrac{2\pi m}{n}\right) & -\sin\left(\dfrac{2\pi m}{n}\right) \\ \sin\left(\dfrac{2\pi m}{n}\right) & \cos\left(\dfrac{2\pi m}{n}\right) \end{bmatrix},$$

which is the matrix for rotation by $2\pi m/n$, and $\psi\colon \mathbb{Z}/n\mathbb{Z} \longrightarrow GL_2(\mathbb{C})$ by

$$\psi_{[m]} = \begin{bmatrix} e^{\frac{2\pi m i}{n}} & 0 \\ 0 & e^{\frac{-2\pi m i}{n}} \end{bmatrix}.$$

Then $\varphi \sim \psi$. To see this, let

$$A = \begin{bmatrix} i & -i \\ 1 & 1 \end{bmatrix},$$

and so

$$A^{-1} = \frac{1}{2i}\begin{bmatrix} 1 & i \\ -1 & i \end{bmatrix}.$$

Then direct computation shows

$$\begin{aligned}
A^{-1}\varphi_{[m]}A &= \frac{1}{2i}\begin{bmatrix} 1 & i \\ -1 & i \end{bmatrix}\begin{bmatrix} \cos\left(\frac{2\pi m}{n}\right) & -\sin\left(\frac{2\pi m}{n}\right) \\ \sin\left(\frac{2\pi m}{n}\right) & \cos\left(\frac{2\pi m}{n}\right) \end{bmatrix}\begin{bmatrix} i & -i \\ 1 & 1 \end{bmatrix} \\
&= \frac{1}{2i}\begin{bmatrix} e^{\frac{2\pi mi}{n}} & ie^{\frac{2\pi mi}{n}} \\ -e^{\frac{-2\pi mi}{n}} & ie^{\frac{-2\pi mi}{n}} \end{bmatrix}\begin{bmatrix} i & -i \\ 1 & 1 \end{bmatrix} \\
&= \frac{1}{2i}\begin{bmatrix} 2ie^{\frac{2\pi mi}{n}} & 0 \\ 0 & 2ie^{\frac{-2\pi mi}{n}} \end{bmatrix} \\
&= \psi_{[m]}.
\end{aligned}$$

The following representation of the symmetric group is very important.

*Example 3.1.9 (Standard representation of $S_n$).* Define $\varphi\colon S_n \longrightarrow GL_n(\mathbb{C})$ on the standard basis by $\varphi_\sigma(e_i) = e_{\sigma(i)}$. One obtains the matrix for $\varphi_\sigma$ by permuting the rows of the identity matrix according to $\sigma$. So, for instance, when $n = 3$ we have

$$\varphi_{(1\ 2)} = \begin{bmatrix} 0 & 1 & 0 \\ 1 & 0 & 0 \\ 0 & 0 & 1 \end{bmatrix}, \quad \varphi_{(1\ 2\ 3)} = \begin{bmatrix} 0 & 0 & 1 \\ 1 & 0 & 0 \\ 0 & 1 & 0 \end{bmatrix}.$$

Notice that in Example 3.1.9

$$\varphi_\sigma(e_1 + e_2 + \cdots + e_n) = e_{\sigma(1)} + e_{\sigma(2)} + \cdots + e_{\sigma(n)} = e_1 + e_2 + \cdots + e_n$$

where the last equality holds since $\sigma$ is a permutation and addition is commutative. Thus $\mathbb{C}(e_1 + \cdots + e_n)$ is invariant under all the $\varphi_\sigma$ with $\sigma \in S_n$. This leads to the following definition.

**Definition 3.1.10 ($G$-invariant subspace).** Let $\varphi\colon G \longrightarrow GL(V)$ be a representation. A subspace $W \leq V$ is *$G$-invariant* if, for all $g \in G$ and $w \in W$, one has $\varphi_g w \in W$.

For $\psi$ from Example 3.1.8, $\mathbb{C}e_1$ and $\mathbb{C}e_2$ are both $\mathbb{Z}/n\mathbb{Z}$-invariant and $\mathbb{C}^2 = \mathbb{C}e_1 \oplus \mathbb{C}e_2$. This is the kind of situation we would like to happen always.

**Definition 3.1.11 (Direct sum of representations).** Suppose that representations $\varphi^{(1)}\colon G \longrightarrow GL(V_1)$ and $\varphi^{(2)}\colon G \longrightarrow GL(V_2)$ are given. Then their (external) *direct sum*

$$\varphi^{(1)} \oplus \varphi^{(2)}\colon G \longrightarrow GL(V_1 \oplus V_2)$$

is given by

$$(\varphi^{(1)} \oplus \varphi^{(2)})_g(v_1, v_2) = (\varphi^{(1)}_g(v_1), \varphi^{(2)}_g(v_2)).$$

Let us try to understand direct sums in terms of matrices. Suppose that $\varphi^{(1)}: G \longrightarrow GL_m(\mathbb{C})$ and $\varphi^{(2)}: G \longrightarrow GL_n(\mathbb{C})$ are representations. Then

$$\varphi^{(1)} \oplus \varphi^{(2)}: G \longrightarrow GL_{m+n}(\mathbb{C})$$

has block matrix form

$$(\varphi^{(1)} \oplus \varphi^{(2)})_g = \begin{bmatrix} \varphi^{(1)}_g & 0 \\ 0 & \varphi^{(2)}_g \end{bmatrix}.$$

*Example 3.1.12.* Define representations $\varphi^{(1)}: \mathbb{Z}/n\mathbb{Z} \longrightarrow \mathbb{C}^*$ by $\varphi^{(1)}_{[m]} = e^{2\pi im/n}$, and $\varphi^{(2)}: \mathbb{Z}/n\mathbb{Z} \longrightarrow \mathbb{C}^*$ by $\varphi^{(2)}_{[m]} = e^{-2\pi im/n}$. Then

$$(\varphi^{(1)} \oplus \varphi^{(2)})_{[m]} = \begin{bmatrix} e^{\frac{2\pi im}{n}} & 0 \\ 0 & e^{\frac{-2\pi im}{n}} \end{bmatrix}.$$

*Remark 3.1.13.* If $n > 1$, then the representation $\rho: G \longrightarrow GL_n(\mathbb{C})$ given by $\rho_g = I$ all $g \in G$ is *not* equivalent to the trivial representation; rather, it is equivalent to the direct sum of $n$ copies of the trivial representation.

Since representations are a special kind of homomorphism, if a group $G$ is generated by a set $X$, then a representation $\varphi$ of $G$ is determined by its values on $X$; of course, not any assignment of matrices to the generators gives a valid representation!

*Example 3.1.14.* Let $\rho: S_3 \longrightarrow GL_2(\mathbb{C})$ be specified on the generators $(1\ 2)$ and $(1\ 2\ 3)$ by

$$\rho_{(1\ 2)} = \begin{bmatrix} -1 & -1 \\ 0 & 1 \end{bmatrix}, \quad \rho_{(1\ 2\ 3)} = \begin{bmatrix} -1 & -1 \\ 1 & 0 \end{bmatrix}$$

(check this is a representation!) and let $\psi: S_3 \longrightarrow \mathbb{C}^*$ be defined by $\psi_\sigma = 1$. Then

$$(\rho \oplus \psi)_{(12)} = \begin{bmatrix} -1 & -1 & 0 \\ 0 & 1 & 0 \\ 0 & 0 & 1 \end{bmatrix}, \quad (\rho \oplus \psi)_{(123)} = \begin{bmatrix} -1 & -1 & 0 \\ 1 & 0 & 0 \\ 0 & 0 & 1 \end{bmatrix}.$$

We shall see later that $\rho \oplus \psi$ is equivalent to the representation of $S_3$ considered in Example 3.1.9.

Let $\varphi: G \longrightarrow GL(V)$ be a representation. If $W \leq V$ is a $G$-invariant subspace, we may restrict $\varphi$ to obtain a representation $\varphi|_W: G \longrightarrow GL(W)$ by setting $(\varphi|_W)_g(w) = \varphi_g(w)$ for $w \in W$. Precisely because $W$ is $G$-invariant, we have $\varphi_g(w) \in W$. Sometime one says $\varphi|_W$ is a *subrepresentation* of $\varphi$. If $V_1, V_2 \leq V$ are $G$-invariant and $V = V_1 \oplus V_2$, then one easily verifies $\varphi$ is equivalent to the

**Table 3.1** Analogies between groups, vector spaces, and representations

| Groups | Vector spaces | Representations |
|---|---|---|
| Subgroup | Subspace | $G$-invariant subspace |
| Simple group | One-dimensional subspace | Irreducible representation |
| Direct product | Direct sum | Direct sum |
| Isomorphism | Isomorphism | Equivalence |

(external) direct sum $\varphi|_{V_1} \oplus \varphi|_{V_2}$. It is instructive to see this in terms of matrices. Let $\varphi^{(i)} = \varphi|_{V_i}$ and choose bases $B_1$ and $B_2$ for $V_1$ and $V_2$, respectively. Then it follows from the definition of a direct sum that $B = B_1 \cup B_2$ is a basis for $V$. Since $V_i$ is $G$-invariant, we have $\varphi_g(B_i) \subseteq V_i = \mathbb{C}B_i$. Thus we have in matrix form

$$[\varphi_g]_B = \begin{bmatrix} [\varphi^{(1)}]_{B_1} & 0 \\ 0 & [\varphi^{(2)}]_{B_2} \end{bmatrix}$$

and so $\varphi \sim \varphi^{(1)} \oplus \varphi^{(2)}$.

In mathematics, it is often the case that one has some sort of unique factorization into primes, or irreducibles. This is the case for representation theory. The notion of "irreducible" in this context is modeled on the notion of a simple group.

**Definition 3.1.15 (Irreducible representation).** A non-zero representation $\varphi\colon G \longrightarrow GL(V)$ of a group $G$ is said to be *irreducible* if the only $G$-invariant subspaces of $V$ are $\{0\}$ and $V$.

*Example 3.1.16.* Any degree one representation $\varphi\colon G \longrightarrow \mathbb{C}^*$ is irreducible, since $\mathbb{C}$ has no proper non-zero subspaces.

Table 3.1 exhibits some analogies between the concepts we have seen so far with ones from Group Theory and Linear Algebra.

If $G = \{1\}$ is the trivial group and $\varphi\colon G \longrightarrow GL(V)$ is a representation, then necessarily $\varphi_1 = I$. So to give a representation of the trivial group is the same thing as to give a vector space. For the trivial group, a $G$-invariant subspace is nothing more than a subspace. A representation of the trivial group is irreducible if and only if it has degree one. So the middle column of Table 3.1 is a special case of the third column.

*Example 3.1.17.* The representations from Example 3.1.8 are not irreducible. For instance,

$$\mathbb{C}\begin{bmatrix} i \\ 1 \end{bmatrix} \quad \text{and} \quad \mathbb{C}\begin{bmatrix} -i \\ 1 \end{bmatrix}$$

are $\mathbb{Z}/n\mathbb{Z}$-invariant subspaces for $\varphi$, while the coordinate axes $\mathbb{C}e_1$ and $\mathbb{C}e_2$ are invariant subspaces for $\psi$.

Not surprisingly, after the one-dimensional representations, the next easiest class to analyze consists of the two-dimensional representations.

*Example 3.1.18.* The representation $\rho\colon S_3 \longrightarrow GL_2(\mathbb{C})$ from Example 3.1.14 is irreducible.

*Proof.* Since $\dim \mathbb{C}^2 = 2$, any non-zero proper $S_3$-invariant subspace $W$ is one-dimensional. Let $v$ be a non-zero vector in $W$; so $W = \mathbb{C}v$. Let $\sigma \in S_3$. Then $\rho_\sigma(v) = \lambda v$ for some $\lambda \in \mathbb{C}$, since by $S_3$-invariance of $W$ we have $\rho_\sigma(v) \in W = \mathbb{C}v$. It follows that $v$ must be an eigenvector for all the $\rho_\sigma$ with $\sigma \in S_3$.

*Claim.* $\rho_{(1\ 2)}$ and $\rho_{(1\ 2\ 3)}$ do not have a common eigenvector.

Indeed, direct computation reveals $\rho_{(1\ 2)}$ has eigenvalues 1 and $-1$ with

$$V_{-1} = \mathbb{C}e_1 \text{ and } V_1 = \mathbb{C}\begin{bmatrix} -1 \\ 2 \end{bmatrix}.$$

Clearly $e_1$ is not an eigenvector of $\rho_{(1\ 2\ 3)}$ as

$$\rho_{(1\ 2\ 3)}\begin{bmatrix} 1 \\ 0 \end{bmatrix} = \begin{bmatrix} -1 \\ 1 \end{bmatrix}.$$

Also,

$$\rho_{(1\ 2\ 3)}\begin{bmatrix} -1 \\ 2 \end{bmatrix} = \begin{bmatrix} -1 \\ -1 \end{bmatrix},$$

so $(-1, 2)$ is not an eigenvector of $\rho_{(1\ 2\ 3)}$. Thus $\rho_{(1\ 2)}$ and $\rho_{(1\ 2\ 3)}$ have no common eigenvector, which implies that $\rho$ is irreducible by the discussion above. □

Let us summarize as a proposition the idea underlying this example.

**Proposition 3.1.19.** *If $\varphi\colon G \longrightarrow GL(V)$ is a representation of degree 2 (i.e., $\dim V = 2$), then $\varphi$ is irreducible if and only if there is no common eigenvector $v$ to all $\varphi_g$ with $g \in G$.*

Notice that this trick of using eigenvectors only works for degree 2 and degree 3 representations (and the latter case requires finiteness of $G$).

*Example 3.1.20.* Let $r$ be rotation by $\pi/2$ and $s$ be reflection over the $x$-axis. These permutations generate the dihedral group $D_4$. Let the representation $\varphi\colon D_4 \longrightarrow GL_2(\mathbb{C})$ be defined by

$$\varphi(r^k) = \begin{bmatrix} i^k & 0 \\ 0 & (-i)^k \end{bmatrix}, \ \varphi(sr^k) = \begin{bmatrix} 0 & (-i)^k \\ i^k & 0 \end{bmatrix}.$$

Then one can apply the previous proposition to check that $\varphi$ is an irreducible representation.

Our eventual goal is to show that each representation is equivalent to a direct sum of irreducible representations. Let us define some terminology to this effect.

**Definition 3.1.21 (Completely reducible).** Let $G$ be a group. A representation $\varphi\colon G \longrightarrow GL(V)$ is said to be *completely reducible* if $V = V_1 \oplus V_2 \oplus \cdots \oplus V_n$ where the $V_i$ are $G$-invariant subspaces and $\varphi|_{V_i}$ is irreducible for all $i = 1, \ldots, n$.

Equivalently, $\varphi$ is completely reducible if $\varphi \sim \varphi^{(1)} \oplus \varphi^{(2)} \oplus \cdots \oplus \varphi^{(n)}$ where the $\varphi^{(i)}$ are irreducible representations.

**Definition 3.1.22 (Decomposable representation).** A non-zero representation $\varphi$ of a group $G$ is *decomposable* if $V = V_1 \oplus V_2$ with $V_1, V_2$ non-zero $G$-invariant subspaces. Otherwise, $V$ is called *indecomposable*.

Complete reducibility is the analog of diagonalizability in representation theory. Our aim is then to show that any representation of a finite group is completely reducible. To do this we show that any representation is either irreducible or decomposable, and then proceed by induction on the degree. First we must show that these notions depend only on the equivalence class of a representation.

**Lemma 3.1.23.** *Let $\varphi\colon G \longrightarrow GL(V)$ be equivalent to a decomposable representation. Then $\varphi$ is decomposable.*

*Proof.* Let $\psi\colon G \longrightarrow GL(W)$ be a decomposable representation with $\psi \sim \varphi$ and $T\colon V \longrightarrow W$ a vector space isomorphism with $\varphi_g = T^{-1}\psi_g T$. Suppose that $W_1$ and $W_2$ are non-zero invariant subspaces of $W$ with $W = W_1 \oplus W_2$. Since $T$ is an equivalence we have that

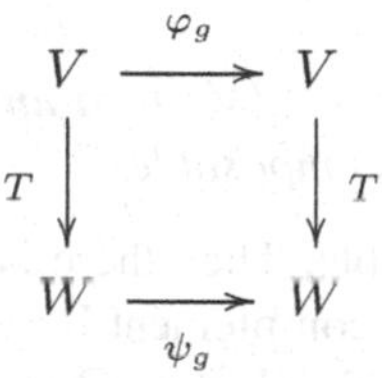

$$\begin{array}{ccc} V & \xrightarrow{\varphi_g} & V \\ {\scriptstyle T}\downarrow & & \downarrow{\scriptstyle T} \\ W & \xrightarrow[\psi_g]{} & W \end{array}$$

commutes, i.e., $T\varphi_g = \psi_g T$ for all $g \in G$. Let $V_1 = T^{-1}(W_1)$ and $V_2 = T^{-1}(W_2)$. First we claim that $V = V_1 \oplus V_2$. Indeed, if $v \in V_1 \cap V_2$, then $Tv \in W_1 \cap W_2 = \{0\}$ and so $Tv = 0$. But $T$ is injective so this implies $v = 0$. Next, if $v \in V$, then $Tv = w_1 + w_2$ some $w_1 \in W_1$ and $w_2 \in W_2$. Then $v = T^{-1}w_1 + T^{-1}w_2 \in V_1 + V_2$. Thus $V = V_1 \oplus V_2$.

Finally, we show that $V_1, V_2$ are $G$-invariant. If $v \in V_i$, then $\varphi_g v = T^{-1}\psi_g T v$. But $Tv \in W_i$ implies $\psi_g T v \in W_i$ since $W_i$ is $G$-invariant. Therefore, we conclude that $\varphi_g v = T^{-1}\psi_g T v \in T^{-1}(W_i) = V_i$, as required. □

We have the analogous results for other types of representations, whose proofs we omit.

**Lemma 3.1.24.** *Let $\varphi\colon G \longrightarrow GL(V)$ be equivalent to an irreducible representation. Then $\varphi$ is irreducible.*

**Lemma 3.1.25.** *Let $\varphi\colon G \longrightarrow GL(V)$ be equivalent to a completely reducible representation. Then $\varphi$ is completely reducible.*

## 3.2 Maschke's Theorem and Complete Reducibility

In order to effect direct sum decompositions of representations, we take advantage of the tools of inner products and orthogonal decompositions.

**Definition 3.2.1 (Unitary representation).** Let $V$ be an inner product space. A representation $\varphi\colon G \longrightarrow GL(V)$ is said to be *unitary* if $\varphi_g$ is unitary for all $g \in G$, i.e.,

$$\langle \varphi_g(v), \varphi_g(w) \rangle = \langle v, w \rangle$$

for all $v, w \in W$. In other words, we may view $\varphi$ as a map $\varphi\colon G \longrightarrow U(V)$.

Identifying $GL_1(\mathbb{C})$ with $\mathbb{C}^*$, we see that a complex number $z$ is unitary (viewed as a matrix) if and only if $\overline{z} = z^{-1}$, that is $z\overline{z} = 1$. But this says exactly that $|z| = 1$, so $U_1(\mathbb{C})$ is exactly the unit circle $\mathbb{T} = \{z \in \mathbb{C} \mid |z| = 1\}$ in $\mathbb{C}$. Hence a one-dimensional unitary representation is a homomorphism $\varphi\colon G \longrightarrow \mathbb{T}$.

*Example 3.2.2.* Define $\varphi\colon \mathbb{R} \longrightarrow \mathbb{T}$ by $\varphi(t) = e^{2\pi it}$. Then $\varphi$ is a unitary representation of the additive group of $\mathbb{R}$ since $\varphi(t+s) = e^{2\pi i(t+s)} = e^{2\pi it}e^{2\pi is} = \varphi(t)\varphi(s)$.

A crucial fact, which makes unitary representations so useful, is that every indecomposable unitary representation is irreducible as the following proposition shows.

**Proposition 3.2.3.** *Let $\varphi\colon G \longrightarrow GL(V)$ be a unitary representation of a group. Then $\varphi$ is either irreducible or decomposable.*

*Proof.* Suppose $\varphi$ is not irreducible. Then there is a non-zero proper $G$-invariant subspace $W$ of $U$. Its orthogonal complement $W^\perp$ is then also non-zero and $V = W \oplus W^\perp$. So it remains to prove that $W^\perp$ is $G$-invariant. If $v \in W^\perp$ and $w \in W$, then

$$\begin{aligned} \langle \varphi_g(v), w \rangle &= \langle \varphi_{g^{-1}}\varphi_g(v), \varphi_{g^{-1}}(w) \rangle && (3.1)\\ &= \langle v, \varphi_{g^{-1}}(w) \rangle && (3.2)\\ &= 0 && (3.3) \end{aligned}$$

where (3.1) follows because $\varphi$ is unitary, (3.2) follows because $\varphi_{g^{-1}}\varphi_g = \varphi_1 = I$ and (3.3) follows because $\varphi_{g^{-1}}w \in W$, as $W$ is $G$-invariant, and $v \in W^\perp$. We conclude $\varphi$ is decomposable. □

It turns out that for finite groups every representation is equivalent to a unitary one. This is not true for infinite groups, as we shall see momentarily.

**Proposition 3.2.4.** *Every representation of a finite group $G$ is equivalent to a unitary representation.*

*Proof.* Let $\varphi\colon G \longrightarrow GL(V)$ be a representation where $\dim V = n$. Choose a basis $B$ for $V$, and let $T\colon V \longrightarrow \mathbb{C}^n$ be the isomorphism taking coordinates with respect to $B$. Then setting $\rho_g = T\varphi_g T^{-1}$, for $g \in G$, yields a representation $\rho\colon G \longrightarrow GL_n(\mathbb{C})$ equivalent to $\varphi$. Let $\langle \cdot, \cdot \rangle$ be the standard inner product on $\mathbb{C}^n$. We define a new inner product $(\cdot, \cdot)$ on $\mathbb{C}^n$ using the crucial "averaging trick." It will be a frequent player throughout the text. Without further ado, define

$$(v, w) = \sum_{g\in G} \langle \rho_g v, \rho_g w \rangle.$$

This summation over $G$, of course, requires that $G$ is finite. It can be viewed as a "smoothing" process.

Let us check that this is indeed an inner product. First we check:

$$\begin{aligned}(c_1v_1 + c_2v_2, w) &= \sum_{g\in G} \langle \rho_g(c_1v_1 + c_2v_2), \rho_g w\rangle \\ &= \sum_{g\in G} \left[c_1\langle \rho_g v_1, \rho_g w\rangle + c_2\langle \rho_g v_2, \rho_g w\rangle\right] \\ &= c_1 \sum_{g\in G} \langle \rho_g v_1, \rho_g w\rangle + c_2 \sum_{g\in G} \langle \rho_g v_2, \rho_g w\rangle \\ &= c_1(v_1, w) + c_2(v_2, w).\end{aligned}$$

Next we verify:

$$\begin{aligned}(w, v) &= \sum_{g\in G} \langle \rho_g w, \rho_g v\rangle \\ &= \sum_{g\in G} \overline{\langle \rho_g v, \rho_g w\rangle} \\ &= \overline{(v, w)}.\end{aligned}$$

Finally, observe that

$$(v, v) = \sum_{g\in G} \langle \rho_g v, \rho_g v\rangle \geq 0$$

because each term $\langle \rho_g v, \rho_g v\rangle \geq 0$. If $(v, v) = 0$, then

$$0 = \sum_{g\in G} \langle \rho_g v, \rho_g v\rangle$$

which implies $\langle \rho_g v, \rho_g v \rangle = 0$ for all $g \in G$ since we are adding non-negative numbers. Hence, $0 = \langle \rho_1 v, \rho_1 v \rangle = \langle v, v \rangle$, and so $v = 0$. We have now established that $(\cdot, \cdot)$ is an inner product.

To verify that the representation is unitary with respect to this inner product, we compute

$$(\rho_h v, \rho_h w) = \sum_{g \in G} \langle \rho_g \rho_h v, \rho_g \rho_h v \rangle = \sum_{g \in G} \langle \rho_{gh} v, \rho_{gh} w \rangle.$$

We now apply a change of variables by setting $x = gh$. As $g$ ranges over all $G$, $x$ ranges over all elements of $G$ since if $k \in G$, then when $g = kh^{-1}$, $x = k$. Therefore,

$$(\rho_h v, \rho_h w) = \sum_{x \in G} \langle \rho_x v, \rho_x w \rangle = (v, w).$$

This completes the proof. □

As a corollary we obtain that every indecomposable representation of a finite group is irreducible.

**Corollary 3.2.5.** *Let $\varphi \colon G \longrightarrow GL(V)$ be a non-zero representation of a finite group. Then $\varphi$ is either irreducible or decomposable.*

*Proof.* By Proposition 3.2.4, $\varphi$ is equivalent to a unitary representation $\rho$. Proposition 3.2.3 then implies that $\rho$ is either irreducible or decomposable. Lemmas 3.1.23 and 3.1.24 then yield that $\varphi$ is either irreducible or decomposable, as was desired. □

The following example shows that Corollary 3.2.5 fails for infinite groups and hence Proposition 3.2.4 must also fail for infinite groups.

*Example 3.2.6.* We provide an example of an indecomposable representation of $\mathbb{Z}$, which is not irreducible. Define $\varphi \colon \mathbb{Z} \longrightarrow GL_2(\mathbb{C})$ by

$$\varphi(n) = \begin{bmatrix} 1 & n \\ 0 & 1 \end{bmatrix}.$$

It is straightforward to verify that $\varphi$ is a homomorphism. The vector $e_1$ is an eigenvector of $\varphi(n)$ for all $n \in \mathbb{Z}$ and so $\mathbb{C}e_1$ is a $\mathbb{Z}$-invariant subspace. This shows that $\varphi$ is not irreducible. On the other hand, if $\varphi$ were decomposable, it would be equivalent to a direct sum of one-dimensional representations. Such a representation is diagonal. But we saw in Example 2.3.5 that $\varphi(1)$ is not diagonalizable. It follows that $\varphi$ is indecomposable.

*Remark 3.2.7.* Observe that any irreducible representation is indecomposable. The previous example shows that the converse fails.

The next theorem is the central result of this chapter. Its proof is quite analogous to the proof of the existence of a prime factorization of an integer or of a factorization of polynomials into irreducibles.

**Theorem 3.2.8 (Maschke).** *Every representation of a finite group is completely reducible.*

*Proof.* Let $\varphi\colon G \longrightarrow GL(V)$ be a representation of a finite group $G$. The proof proceeds by induction on the degree of $\varphi$, that is, $\dim V$. If $\dim V = 1$, then $\varphi$ is irreducible since $V$ has no non-zero proper subspaces. Assume the statement is true for $\dim V \leq n$. Let $\varphi\colon G \longrightarrow GL(V)$ be a representation with $\dim V = n+1$. If $\varphi$ is irreducible, then we are done. Otherwise, $\varphi$ is decomposable by Corollary 3.2.5, and so $V = V_1 \oplus V_2$ where $0 \neq V_1, V_2$ are $G$-invariant subspaces. Since $\dim V_1, \dim V_2 < \dim V$, by induction, $\varphi|_{V_1}$ and $\varphi|_{V_2}$ are completely reducible. Therefore, $V_1 = U_1 \oplus \cdots \oplus U_s$ and $V_2 = W_1 \oplus \cdots \oplus W_r$ where the $U_i, W_j$ are $G$-invariant and the subrepresentations $\varphi|_{U_i}, \varphi|_{W_j}$ are irreducible for all $1 \leq i \leq s, 1 \leq j \leq r$. Then $V = U_1 \oplus \cdots U_s \oplus W_1 \oplus \cdots \oplus W_r$ and hence $\varphi$ is completely irreducible. □

*Remark 3.2.9.* If one follows the details of the proof carefully, one can verify that if $\varphi$ is a unitary matrix representation, then $\varphi$ is equivalent to a direct sum of irreducible unitary representations via an equivalence implemented by a unitary matrix $T$.

In conclusion, if $\varphi\colon G \longrightarrow GL_n(\mathbb{C})$ is any representation of a finite group, then

$$\varphi \sim \begin{bmatrix} \varphi^{(1)} & 0 & \cdots & 0 \\ 0 & \varphi^{(2)} & \ddots & \vdots \\ \vdots & \ddots & \ddots & 0 \\ 0 & \cdots & 0 & \varphi^{(m)} \end{bmatrix}$$

where the $\varphi^{(i)}$ are irreducible for all $i$. This is analogous to the spectral theorem stating that all self-adjoint matrices are diagonalizable.

There still remains the question as to whether the decomposition into irreducible representations is unique. This will be resolved in the next chapter.

## Exercises

**Exercise 3.1.** Let $\varphi\colon D_4 \longrightarrow \mathrm{GL}_2(\mathbb{C})$ be the representation given by

$$\varphi(r^k) = \begin{bmatrix} i^k & 0 \\ 0 & (-i)^k \end{bmatrix}, \varphi(sr^k) = \begin{bmatrix} 0 & (-i)^k \\ i^k & 0 \end{bmatrix}$$

where $r$ is rotation counterclockwise by $\pi/2$ and $s$ is reflection over the $x$-axis. Prove that $\varphi$ is irreducible.

**Exercise 3.2.** Prove Lemma 3.1.24.

**Exercise 3.3.** Let $\varphi, \psi \colon G \longrightarrow \mathbb{C}^*$ be one-dimensional representations. Show that $\varphi$ is equivalent to $\psi$ if and only if $\varphi = \psi$.

**Exercise 3.4.** Let $\varphi \colon G \longrightarrow \mathbb{C}^*$ be a representation. Suppose that $g \in G$ has order $n$.

1. Show that $\varphi(g)$ is an $n$th-root of unity (i.e., a solution to the equation $z^n = 1$).
2. Construct $n$ inequivalent one-dimensional representations $\mathbb{Z}/n\mathbb{Z} \longrightarrow \mathbb{C}^*$.
3. Explain why your representations are the only possible one-dimensional representations.

**Exercise 3.5.** Let $\varphi \colon G \longrightarrow GL(V)$ be a representation of a finite group $G$. Define the *fixed subspace*

$$V^G = \{v \in V \mid \varphi_g v = v, \forall g \in G\}.$$

1. Show that $V^G$ is a $G$-invariant subspace.
2. Show that
$$\frac{1}{|G|} \sum_{h \in G} \varphi_h v \in V^G$$
for all $v \in V$.
3. Show that if $v \in V^G$, then
$$\frac{1}{|G|} \sum_{h \in G} \varphi_h v = v.$$
4. Conclude $\dim V^G$ is the rank of the operator
$$P = \frac{1}{|G|} \sum_{h \in G} \varphi_h.$$
5. Show that $P^2 = P$.
6. Conclude $\mathrm{Tr}(P)$ is the rank of $P$.
7. Conclude
$$\dim V^G = \frac{1}{|G|} \sum_{h \in G} \mathrm{Tr}(\varphi_h).$$

**Exercise 3.6.** Let $\varphi \colon G \longrightarrow GL_n(\mathbb{C})$ be a representation.

1. Show that setting $\psi_g = \overline{\varphi_g}$ provides a representation $\psi \colon G \longrightarrow GL_n(\mathbb{C})$. It is called the *conjugate representation*. Give an example showing that $\varphi$ and $\psi$ do not have to be equivalent.
2. Let $\chi \colon G \longrightarrow \mathbb{C}^*$ be a degree 1 representation of $G$. Define a map $\varphi^\chi \colon G \longrightarrow GL_n(\mathbb{C})$ by $\varphi^\chi_g = \chi(g)\varphi_g$. Show that $\varphi^\chi$ is a representation. Give an example showing that $\varphi$ and $\varphi^\chi$ do not have to be equivalent.

**Exercise 3.7.** Give a bijection between unitary, degree one representations of $\mathbb{Z}$ and elements of $\mathbb{T}$.

**Exercise 3.8.**

1. Let $\varphi\colon G \longrightarrow GL_3(\mathbb{C})$ be a representation of a finite group. Show that $\varphi$ is irreducible if and only if there is no common eigenvector for the matrices $\varphi_g$ with $g \in G$.
2. Given an example of a finite group $G$ and a decomposable representation $\varphi\colon G \longrightarrow GL_4(\mathbb{C})$ such that the $\varphi_g$ with $g \in G$ do not have a common eigenvector.

# Chapter 4
# Character Theory and the Orthogonality Relations

This chapter gets to the heart of group representation theory: the character theory of Frobenius and Schur. The fundamental idea of character theory is to encode a representation $\varphi\colon G \longrightarrow GL_n(\mathbb{C})$ of $G$ by a complex-valued function $\chi_\varphi\colon G \longrightarrow C$. In other words, we replace a function to an $n$-dimensional space with a function to a 1-dimensional space. The characters turn out to form an orthonormal set with respect to the inner product on functions, a fact that we shall use to prove the uniqueness of the decomposition of a representation into irreducibles.

Much of the chapter is dedicated to establishing the Schur orthogonality relations, which roughly speaking says that the entries of the irreducible unitary representations of a finite group $G$ form an orthogonal basis for the space of complex-valued functions on $G$. In the next chapter, we use these results to develop Fourier analysis on finite groups, which is central to applications.

## 4.1 Morphisms of Representations

A principle of modern mathematics is that the maps between mathematical objects should be placed on equal footing with the objects themselves. Keeping this in mind, we turn to the notion of "morphism" between representations. The idea is the following. Let $\varphi\colon G \longrightarrow GL(V)$ be a representation. We can think of elements of $G$ as scalars via $g \cdot v = \varphi_g v$ for $v \in V$. A morphism between $\varphi\colon G \longrightarrow GL(V)$ and $\rho\colon G \longrightarrow GL(W)$ should be a linear transformation $T\colon V \longrightarrow W$ such that $Tgv = gTv$ for all $g \in G$ and $v \in V$. Formally, this means $T\varphi_g v = \rho_g Tv$ for all $v \in V$, i.e., $T\varphi_g = \rho_g T$ for all $g \in G$.

**Definition 4.1.1 (Morphism).** Let $\varphi\colon G \longrightarrow GL(V)$ and $\rho\colon G \longrightarrow GL(W)$ be representations. A *morphism*[1] from $\varphi$ to $\rho$ is by definition a linear map $T\colon V \longrightarrow W$ such that $T\varphi_g = \rho_g T$, for all $g \in G$. In other words the diagram

[1] Some authors use the term *intertwiner* or *intertwining operator* for what we call morphism.

B. Steinberg, *Representation Theory of Finite Groups: An Introductory Approach*, Universitext, DOI 10.1007/978-1-4614-0776-8_4,
© Springer Science+Business Media, LLC 2012

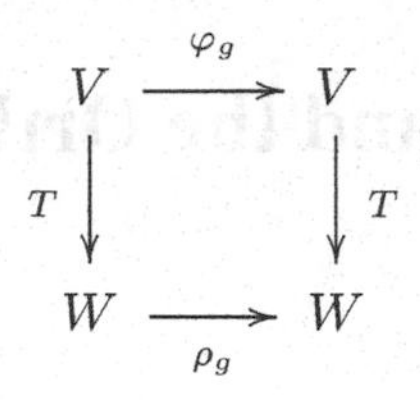


commutes for all $g \in G$.

The set of all morphisms from $\varphi$ to $\rho$ is denoted $\mathrm{Hom}_G(\varphi, \rho)$. Notice that $\mathrm{Hom}_G(\varphi, \rho) \subseteq \mathrm{Hom}(V, W)$.

*Remark 4.1.2.* If $T \in \mathrm{Hom}_G(\varphi, \rho)$ is invertible, then $\varphi \sim \rho$ and $T$ is an *equivalence* (or *isomorphism*).

*Remark 4.1.3.* Observe that $T\colon V \longrightarrow V$ belongs to $\mathrm{Hom}_G(\varphi, \varphi)$ if and only if $T\varphi_g = \varphi_g T$ for all $g \in G$, i.e., $T$ commutes with (or centralizes) $\varphi(G)$. In particular, the identity map $I\colon V \longrightarrow V$ is always an element of $\mathrm{Hom}_G(\varphi, \varphi)$.

As is typical for homomorphisms in algebra, the kernel and the image of a morphism of representations are subrepresentations.

**Proposition 4.1.4.** *Let* $T\colon V \longrightarrow W$ *be in* $\mathrm{Hom}_G(\varphi, \rho)$. *Then* $\ker T$ *is a* $G$*-invariant subspace of* $V$ *and* $T(V) = \mathrm{Im}\, T$ *is a* $G$*-invariant subspace of* $W$.

*Proof.* Let $v \in \ker T$ and $g \in G$. Then $T\varphi_g v = \rho_g T v = 0$ since $v \in \ker T$. Hence $\varphi_g v \in \ker T$. We conclude that $\ker T$ is $G$-invariant.

Now let $w \in \mathrm{Im}\, T$, say $w = Tv$ with $v \in V$. Then $\rho_g w = \rho_g T v = T\varphi_g v \in \mathrm{Im}\, T$, establishing that $\mathrm{Im}\, T$ is $G$-invariant. □

The set of morphisms from $\varphi$ to $\rho$ has the additional structure of a vector space, as the following proposition reveals.

**Proposition 4.1.5.** *Let* $\varphi\colon G \longrightarrow GL(V)$ *and* $\rho\colon G \longrightarrow GL(W)$ *be representations. Then* $\mathrm{Hom}_G(\varphi, \rho)$ *is a subspace of* $\mathrm{Hom}(V, W)$.

*Proof.* Let $T_1, T_2 \in \mathrm{Hom}_G(\varphi, \rho)$ and $c_1, c_2 \in \mathbb{C}$. Then

$$(c_1T_1 + c_2T_2)\varphi_g = c_1T_1\varphi_g + c_2T_2\varphi_g = c_1\rho_gT_1 + c_2\rho_gT_2 = \rho_g(c_1T_1 + c_2T_2)$$

and hence $c_1T_1 + c_2T_2 \in \mathrm{Hom}_G(\varphi, \rho)$, as required. □

Fundamental to all of representation theory is the important observation, due to I. Schur, that roughly speaking morphisms between irreducible representations are very limited. This is the first place where we seriously use that we are working over the field of complex numbers and not the field of real numbers. Namely, we use that every linear operator on a finite-dimensional complex vector space has an eigenvalue. This is a consequence of the fact that every polynomial over $\mathbb{C}$ has a root; in particular the characteristic polynomial of the operator has a root.

**Lemma 4.1.6 (Schur's lemma).** *Let $\varphi, \rho$ be irreducible representations of $G$, and $T \in \mathrm{Hom}_G(\varphi, \rho)$. Then either $T$ is invertible or $T = 0$. Consequently:*

*(a) If $\varphi \not\sim \rho$, then $\mathrm{Hom}_G(\varphi, \rho) = 0$;*
*(b) If $\varphi = \rho$, then $T = \lambda I$ with $\lambda \in \mathbb{C}$ (i.e., $T$ is multiplication by a scalar).*

*Proof.* Let $\varphi\colon G \longrightarrow GL(V)$, $\rho\colon G \longrightarrow GL(W)$, and let $T\colon V \longrightarrow W$ be in $\mathrm{Hom}_G(\varphi, \rho)$. If $T = 0$, we are done; so assume that $T \neq 0$. Proposition 4.1.4 implies that $\ker T$ is $G$-invariant and hence either $\ker T = V$ or $\ker T = 0$. Since $T \neq 0$, the former does not happen; thus $\ker T = 0$ and so $T$ is injective. Also, according to Proposition 4.1.4, $\mathrm{Im}\, T$ is $G$-invariant, so $\mathrm{Im}\, T = W$ or $\mathrm{Im}\, T = 0$. If $\mathrm{Im}\, T = 0$, then again $T = 0$. So it must be $\mathrm{Im}\, T = W$, that is, $T$ is surjective. We conclude that $T$ is invertible.

For (a), assume $\mathrm{Hom}_G(\varphi, \rho) \neq 0$. That means there exists $T \neq 0$ in $\mathrm{Hom}_G(\varphi, \rho)$. Then $T$ is invertible, by the above, and so $\varphi \sim \rho$. This is the contrapositive of what we wanted to show.

To establish (b), let $\lambda$ be an eigenvalue of $T$ (here is where we use that we are working over $\mathbb{C}$ and not $\mathbb{R}$). Then $\lambda I - T$ is not invertible by definition of an eigenvalue. As $I \in \mathrm{Hom}_G(\varphi, \varphi)$, Proposition 4.1.5 tells us that $\lambda I - T$ belongs to $\mathrm{Hom}_G(\varphi, \varphi)$. Since all non-zero elements of $\mathrm{Hom}_G(\varphi, \varphi)$ are invertible by the first paragraph of the proof, it follows $\lambda I - T = 0$. Of course, this is the same as saying $T = \lambda I$. □

*Remark 4.1.7.* It is not hard to deduce from Schur's lemma that if $\varphi$ and $\rho$ are equivalent irreducible representations, then $\dim \mathrm{Hom}_G(\varphi, \rho) = 1$.

We are now in a position to describe the irreducible representations of an abelian group.

**Corollary 4.1.8.** *Let $G$ be an abelian group. Then any irreducible representation of $G$ has degree one.*

*Proof.* Let $\varphi\colon G \longrightarrow GL(V)$ be an irreducible representation. Fix for the moment $h \in G$. Then setting $T = \varphi_h$, we obtain, for all $g \in G$, that

$$T\varphi_g = \varphi_h\varphi_g = \varphi_{hg} = \varphi_{gh} = \varphi_g\varphi_h = \varphi_g T.$$

Consequently, Schur's lemma implies $\varphi_h = \lambda_h I$ for some scalar $\lambda_h \in \mathbb{C}$ (the subscript indicates the dependence on $h$). Let $v$ be a non-zero vector in $V$ and $k \in \mathbb{C}$. Then $\varphi_h(kv) = \lambda_h Ikv = \lambda_h kv \in \mathbb{C}v$. Thus $\mathbb{C}v$ is a $G$-invariant subspace, as $h$ was arbitrary. We conclude that $V = \mathbb{C}v$ by irreducibility and so $\dim V = 1$. □

Let us present some applications of this result to linear algebra.

**Corollary 4.1.9.** *Let $G$ be a finite abelian group and $\varphi\colon G \longrightarrow GL_n(\mathbb{C})$ a representation. Then there is an invertible matrix $T$ such that $T^{-1}\varphi_g T$ is diagonal for all $g \in G$ ($T$ is independent of $g$).*

*Proof.* Since $\varphi$ is completely reducible, we have that $\varphi \sim \varphi^{(1)} \oplus \cdots \oplus \varphi^{(m)}$ where $\varphi^{(1)}, \ldots, \varphi^{(m)}$ are irreducible. Since $G$ is abelian, the degree of each $\varphi^{(i)}$ is 1 (and hence $n = m$). Consequently, $\varphi_g^{(i)} \in \mathbb{C}^*$ for all $g \in G$. Now if $T\colon \mathbb{C}^n \longrightarrow \mathbb{C}^n$ gives the equivalence of $\varphi$ with $\varphi^{(1)} \oplus \cdots \oplus \varphi^{(n)}$, then

$$T^{-1}\varphi_g T = \begin{bmatrix} \varphi_g^{(1)} & 0 & \cdots & 0 \\ 0 & \varphi_g^{(2)} & \ddots & \vdots \\ \vdots & \ddots & \ddots & 0 \\ 0 & \cdots & 0 & \varphi_g^{(n)} \end{bmatrix}$$

is diagonal for all $g \in G$. □

As a corollary, we obtain the diagonalizability of matrices of finite order.

**Corollary 4.1.10.** *Let $A \in GL_m(\mathbb{C})$ be a matrix of finite order. Then $A$ is diagonalizable. Moreover, if $A^n = I$, then the eigenvalues of $A$ are $n$th-roots of unity.*

*Proof.* Suppose that $A^n = I$. Define a representation $\varphi\colon \mathbb{Z}/n\mathbb{Z} \longrightarrow GL_m(\mathbb{C})$ by setting $\varphi([k]) = A^k$. This is easily verified to give a well-defined representation. Thus there exists $T \in GL_n(\mathbb{C})$ such that $T^{-1}AT$ is diagonal by Corollary 4.1.9. Suppose

$$T^{-1}AT = \begin{bmatrix} \lambda_1 & 0 & \cdots & 0 \\ 0 & \lambda_2 & \ddots & \vdots \\ \vdots & \ddots & \ddots & 0 \\ 0 & \cdots & 0 & \lambda_m \end{bmatrix} = D.$$

Then

$$D^n = (T^{-1}AT)^n = T^{-1}A^nT = T^{-1}IT = I.$$

Therefore, we have

$$\begin{bmatrix} \lambda_1^n & 0 & \cdots & 0 \\ 0 & \lambda_2^n & \ddots & \vdots \\ \vdots & \ddots & \ddots & 0 \\ 0 & \cdots & 0 & \lambda_m^n \end{bmatrix} = D^n = I$$

and so $\lambda_i^n = 1$ for all $i$. This establishes that the eigenvalues of $A$ are $n$th-roots of unity. □

*Remark 4.1.11.* Corollary 4.1.10 can also be proved easily using minimal polynomials.

## 4.2 The Orthogonality Relations

From this point onward, the group $G$ shall always be assumed finite. Let $\varphi\colon G \longrightarrow GL_n(\mathbb{C})$ be a representation. Then $\varphi_g = (\varphi_{ij}(g))$ where $\varphi_{ij}(g) \in \mathbb{C}$ for $1 \leq i, j \leq n$. Thus there are $n^2$ functions $\varphi_{ij}\colon G \longrightarrow \mathbb{C}$ associated to the degree $n$ representation $\varphi$. What can be said about the functions $\varphi_{ij}$ when $\varphi$ is irreducible and unitary? It turns out that the functions of this sort form an orthogonal basis for $\mathbb{C}^G$.

**Definition 4.2.1 (Group algebra).** Let $G$ be a group and define

$$L(G) = \mathbb{C}^G = \{f \mid f\colon G \longrightarrow \mathbb{C}\}.$$

Then $L(G)$ is an inner product space with addition and scalar multiplication given by

$$(f_1 + f_2)(g) = f_1(g) + f_2(g)$$
$$(cf)(g) = c \cdot f(g)$$

and with the inner product defined by

$$\langle f_1, f_2 \rangle = \frac{1}{|G|} \sum_{g \in G} f_1(g)\overline{f_2(g)}.$$

For reasons to become apparent later, $L(G)$ is called the *group algebra* of $G$.

One of our goals in this chapter is to prove the following important result due to I. Schur. Recall that $U_n(\mathbb{C})$ is the group of $n \times n$ unitary matrices.

**Theorem (Schur orthogonality relations).** *Suppose that $\varphi\colon G \longrightarrow U_n(\mathbb{C})$ and $\rho\colon G \longrightarrow U_m(\mathbb{C})$ are inequivalent irreducible unitary representations. Then:*

1. $\langle \varphi_{ij}, \rho_{k\ell} \rangle = 0$;
2. $\langle \varphi_{ij}, \varphi_{k\ell} \rangle = \begin{cases} 1/n & \text{if } i = k \text{ and } j = \ell \\ 0 & \text{else.} \end{cases}$

The proof requires a lot of preparation. We begin with our second usage of the "averaging trick."

**Proposition 4.2.2.** *Let $\varphi\colon G \longrightarrow GL(V)$ and $\rho\colon G \longrightarrow GL(W)$ be representations and suppose that $T\colon V \longrightarrow W$ is a linear transformation. Then:*

*(a)* $T^\sharp = \frac{1}{|G|}\sum_{g \in G} \rho_{g^{-1}} T \varphi_g \in \mathrm{Hom}_G(\varphi, \rho)$
*(b)* *If* $T \in \mathrm{Hom}_G(\varphi, \rho)$, *then* $T^\sharp = T$.
*(c)* *The map* $P\colon \mathrm{Hom}(V, W) \longrightarrow \mathrm{Hom}_G(\varphi, \rho)$ *defined by* $P(T) = T^\sharp$ *is an onto linear map.*

*Proof.* We verify (a) by a direct computation.

$$T^{\sharp}\varphi_h = \frac{1}{|G|}\sum_{g\in G}\rho_{g^{-1}}T\varphi_g\varphi_h = \frac{1}{|G|}\sum_{g\in G}\rho_{g^{-1}}T\varphi_{gh}. \tag{4.1}$$

The next step is to apply a change of variables $x = gh$. Since right multiplication by $h$ is a permutation of $G$, as $g$ varies over $G$, so does $x$. Noting that $g^{-1} = hx^{-1}$, we conclude that the right-hand side of (4.1) is equal to

$$\frac{1}{|G|}\sum_{x\in G}\rho_{hx^{-1}}T\varphi_x = \frac{1}{|G|}\sum_{x\in G}\rho_h\rho_{x^{-1}}T\varphi_x = \rho_h\frac{1}{|G|}\sum_{x\in G}\rho_{x^{-1}}T\varphi_x = \rho_hT^{\sharp}.$$

This proves $T^{\sharp}\in \mathrm{Hom}_G(\varphi,\rho)$.

To prove (b), notice that if $T\in \mathrm{Hom}_G(\varphi,\rho)$, then

$$T^{\sharp} = \frac{1}{|G|}\sum_{g\in G}\rho_{g^{-1}}T\varphi_g = \frac{1}{|G|}\sum_{g\in G}\rho_{g^{-1}}\rho_gT = \frac{1}{|G|}\sum_{g\in G}T = \frac{1}{|G|}|G|T = T.$$

Finally, for (c) we establish linearity by checking

$$\begin{aligned} P(c_1T_1+c_2T_2) &= (c_1T_1+c_2T_2)^{\sharp} \\ &= \frac{1}{|G|}\sum_{g\in G}\rho_{g^{-1}}(c_1T_1+c_2T_2)\varphi_g \\ &= c_1\frac{1}{|G|}\sum_{g\in G}\rho_{g^{-1}}T_1\varphi_g + c_2\frac{1}{|G|}\sum_{g\in G}\rho_{g^{-1}}T_2\varphi_g \\ &= c_1T_1^{\sharp}+c_2T_2^{\sharp} = c_1P(T_1)+c_2P(T_2). \end{aligned}$$

If $T\in \mathrm{Hom}_G(\varphi,\rho)$, then (b) implies $T = T^{\sharp} = P(T)$ and so $P$ is onto. □

The following variant of Schur's lemma will be the form in which we shall most commonly use it. It is based on the trivial observation that if $I_n$ is the $n\times n$ identity matrix and $\lambda\in\mathbb{C}$, then $\mathrm{Tr}(\lambda I_n) = n\lambda$.

**Proposition 4.2.3.** *Let $\varphi\colon G\longrightarrow GL(V)$, $\rho\colon G\longrightarrow GL(W)$ be irreducible representations of $G$ and let $T\colon V\longrightarrow W$ be a linear map. Then:*

*(a) If $\varphi\nsim\rho$, then $T^{\sharp}=0$;*
*(b) If $\varphi=\rho$, then $T^{\sharp} = \frac{\mathrm{Tr}(T)}{\deg\varphi}I$.*

*Proof.* Assume first $\varphi\nsim\rho$. Then $\mathrm{Hom}_G(\varphi,\rho) = 0$ by Schur's lemma and so $T^{\sharp}=0$. Next suppose $\varphi=\rho$. By Schur's lemma, $T^{\sharp}=\lambda I$ some $\lambda\in\mathbb{C}$. Our goal is to solve for $\lambda$. As $T^\sharp\colon V \longrightarrow V$, we have $\mathrm{Tr}(\lambda I) = \lambda \mathrm{Tr}(I) = \lambda \dim V = \lambda \deg \varphi$. It follows that $T^\sharp = \frac{\mathrm{Tr}(T^\sharp)}{\deg \varphi} I$.

On the other hand, we can also compute the trace directly from the definition of $T^\sharp$. Using $\mathrm{Tr}(AB) = \mathrm{Tr}(BA)$, we obtain

$$\mathrm{Tr}(T^\sharp) = \frac{1}{|G|}\sum_{g\in G}\mathrm{Tr}(\varphi_{g^{-1}}T\varphi_g) = \frac{1}{|G|}\sum_{g\in G}\mathrm{Tr}(T) = \frac{|G|}{|G|}\mathrm{Tr}(T) = \mathrm{Tr}(T)$$

and so $T^\sharp = \frac{\mathrm{Tr}(T)}{\deg \varphi} I$, as required. □

If $\varphi\colon G \longrightarrow GL_n(\mathbb{C})$ and $\rho\colon G \longrightarrow GL_m(\mathbb{C})$ are representations, then $\mathrm{Hom}(V, W) = M_{mn}(\mathbb{C})$ and $\mathrm{Hom}_G(\varphi, \rho)$ is a subspace of $M_{mn}(\mathbb{C})$. Hence the map $P$ from Proposition 4.2.2 can be viewed as a linear transformation $P\colon M_{mn}(\mathbb{C}) \longrightarrow M_{mn}(\mathbb{C})$. It would then be natural to compute the matrix of $P$ with respect to the standard basis for $M_{mn}(\mathbb{C})$. It turns out that when $\varphi$ and $\rho$ are unitary representations, the matrix for $P$ has a special form. Recall that the standard basis for $M_{mn}(\mathbb{C})$ consists of the matrices $E_{11}, E_{12}, \ldots, E_{mn}$ where $E_{ij}$ is the $m \times n$-matrix with 1 in position $ij$ and 0 elsewhere. One then has $(a_{ij}) = \sum_{ij} a_{ij}E_{ij}$.

The following lemma is a straightforward computation with the formula for matrix multiplication.

**Lemma 4.2.4.** *Let $A \in M_{rm}(\mathbb{C})$, $B \in M_{ns}(\mathbb{C})$ and $E_{ki} \in M_{mn}(\mathbb{C})$. Then the formula $(AE_{ki}B)_{\ell j} = a_{\ell k}b_{ij}$ holds where $A = (a_{ij})$ and $B = (b_{ij})$.*

*Proof.* By definition

$$(AE_{ki}B)_{\ell j} = \sum_{x,y} a_{\ell x}(E_{ki})_{xy}b_{yj}.$$

But all terms in this sum are 0, except when $x = k$, $y = i$, in which case one obtains $a_{\ell k}b_{ij}$, as desired. □

*Example 4.2.5.* This example illustrates Lemma 4.2.4:

$$\begin{bmatrix} a_{11} & a_{12} \\ a_{21} & a_{22} \end{bmatrix}\begin{bmatrix} 0 & 1 \\ 0 & 0 \end{bmatrix}\begin{bmatrix} b_{11} & b_{12} \\ b_{21} & b_{22} \end{bmatrix} = \begin{bmatrix} 0 & a_{11} \\ 0 & a_{21} \end{bmatrix}\begin{bmatrix} b_{11} & b_{12} \\ b_{21} & b_{22} \end{bmatrix} = \begin{bmatrix} a_{11}b_{21} & a_{11}b_{22} \\ a_{21}b_{21} & a_{21}b_{22} \end{bmatrix}.$$

Now we are prepared to compute the matrix of $P$ with respect to the standard basis. We state the result in the form in which we shall use it.

**Lemma 4.2.6.** *Let $\varphi\colon G \longrightarrow U_n(\mathbb{C})$ and $\rho\colon G \longrightarrow U_m(\mathbb{C})$ be unitary representations. Let $A = E_{ki} \in M_{mn}(\mathbb{C})$. Then $A^\sharp_{\ell j} = \langle \varphi_{ij}, \rho_{k\ell}\rangle$.*

*Proof.* Since $\rho$ is unitary, $\rho_{g^{-1}} = \rho_g^{-1} = \rho_g^*$. Thus $\rho_{\ell k}(g^{-1}) = \overline{\rho_{k\ell}(g)}$. Keeping this in mind, we compute

$$\begin{aligned} A^{\sharp}_{\ell j} &= \frac{1}{|G|}\sum_{g\in G}(\rho_{g^{-1}}E_{ki}\varphi_g)_{\ell j} \\ &= \frac{1}{|G|}\sum_{g\in G}\rho_{\ell k}(g^{-1})\varphi_{ij}(g) \qquad \text{by Lemma 4.2.4} \\ &= \frac{1}{|G|}\sum_{g\in G}\overline{\rho_{k\ell}(g)}\varphi_{ij}(g) \\ &= \langle \varphi_{ij}, \rho_{k\ell}\rangle \end{aligned}$$

as required. □

*Remark 4.2.7.* Let $P\colon M_{mn}(\mathbb{C}) \longrightarrow M_{mn}(\mathbb{C})$ be the linear transformation given by $P(T) = T^{\sharp}$ and let $B$ be the matrix of $P$ with respect to the standard basis for $M_{mn}(\mathbb{C})$. Then $B$ is an $mn \times mn$ matrix whose rows and columns are indexed by pairs $\ell j, ki$ where $1 \leq \ell, k \leq m$ and $1 \leq j, i \leq n$. The content of Lemma 4.2.6 is that the $\ell j, ki$ entry of $B$ is the inner product $\langle \varphi_{ij}, \rho_{k\ell}\rangle$.

We can now prove the Schur orthogonality relations.

**Theorem 4.2.8 (Schur orthogonality relations).** *Let $\varphi\colon G \longrightarrow U_n(\mathbb{C})$ and $\rho\colon G \longrightarrow U_m(\mathbb{C})$ be inequivalent irreducible unitary representations. Then:*

1. $\langle \varphi_{ij}, \rho_{k\ell}\rangle = 0$;
2. $\langle \varphi_{ij}, \varphi_{k\ell}\rangle = \begin{cases} 1/n & \text{if } i = k \text{ and } j = \ell \\ 0 & \text{else.}\end{cases}$

*Proof.* For 1, let $A = E_{ki} \in M_{mn}(\mathbb{C})$. Then $A^{\sharp} = 0$ by Proposition 4.2.3. On the other hand, $A^{\sharp}_{\ell j} = \langle \varphi_{ij}, \rho_{k\ell}\rangle$ by Lemma 4.2.6. This establishes 1.

Next, we apply Proposition 4.2.3 and Lemma 4.2.6 with $\varphi = \rho$. Let $A = E_{ki} \in M_n(\mathbb{C})$. Then

$$A^{\sharp} = \frac{\operatorname{Tr}(E_{ki})}{n} I$$

by Proposition 4.2.3. Lemma 4.2.6 shows that $A^{\sharp}_{\ell j} = \langle \varphi_{ij}, \varphi_{k\ell}\rangle$. First suppose that $j \neq \ell$. Then since $I_{\ell j} = 0$, it follows $0 = A^{\sharp}_{\ell j} = \langle \varphi_{ij}, \varphi_{k\ell}\rangle$. Next suppose that $i \neq k$. Then $E_{ki}$ has only zeroes on the diagonal and so $\operatorname{Tr}(E_{ki}) = 0$. Thus we again have $0 = A^{\sharp}_{\ell j} = \langle \varphi_{ij}, \varphi_{k\ell}\rangle$. Finally, in the case where $\ell = j$ and $i = k$, $E_{ki}$ has a single 1 on the diagonal and all other entries are 0. Thus $\operatorname{Tr}(E_{ki}) = 1$ and so $1/n = A^{\sharp}_{\ell j} = \langle \varphi_{ij}, \varphi_{k\ell}\rangle$. This proves the theorem. □

A simple renormalization establishes:

**Corollary 4.2.9.** *Let $\varphi$ be an irreducible unitary representation of $G$ of degree $d$. Then the $d^2$ functions $\{\sqrt{d}\varphi_{ij} \mid 1 \leq i, j \leq d\}$ form an orthonormal set.*

An important corollary of Theorem 4.2.8 is that there are only finitely many equivalence classes of irreducible representations of $G$. First recall that every equivalence class contains a unitary representation. Next, because $\dim L(G) = |G|$, no linearly independent set of vectors from $L(G)$ can have more than $|G|$ elements. Theorem 4.2.8 says that the entries of inequivalent unitary representations of $G$ form an orthogonal set of non-zero vectors in $L(G)$. It follows that $G$ has at most $|G|$ equivalence classes of irreducible representations. In fact, if $\varphi^{(1)}, \ldots, \varphi^{(s)}$ are a complete set of representatives of the equivalence classes of irreducible representations of $G$ and $d_i = \deg \varphi^{(i)}$, then the $d_1^2 + d_2^2 + \cdots + d_s^2$ functions $\{\sqrt{d_k}\varphi_{ij}^{(k)} \mid 1 \leq k \leq s, 1 \leq i, j \leq d_k\}$ form an orthonormal set of vectors in $L(G)$ and hence $s \leq d_1^2 + \cdots + d_s^2 \leq |G|$ (the first inequality holds since $d_i \geq 1$ all $i$). We summarize this discussion in the following proposition.

**Proposition 4.2.10.** *Let $G$ be a finite group. Let $\varphi^{(1)}, \ldots, \varphi^{(s)}$ be a complete set of representatives of the equivalence classes of irreducible representations of $G$ and set $d_i = \deg \varphi^{(i)}$. Then the functions*

$$\left\{\sqrt{d_k}\varphi_{ij}^{(k)} \mid 1 \leq k \leq s, 1 \leq i, j \leq d_k\right\}$$

*form an orthonormal set in $L(G)$ and hence $s \leq d_1^2 + \cdots + d_s^2 \leq |G|$.*

Later, we shall see that the second inequality in the proposition is in fact an equality; the first one is only an equality for abelian groups.

## 4.3 Characters and Class Functions

In this section, we finally prove the uniqueness of the decomposition of a representation into irreducible representations. The key ingredient is to associate to each representation $\varphi$ a function $\chi_\varphi \colon G \longrightarrow \mathbb{C}$ which encodes the entire representation.

**Definition 4.3.1 (Character).** Let $\varphi \colon G \longrightarrow GL(V)$ be a representation. The *character* $\chi_\varphi \colon G \longrightarrow \mathbb{C}$ of $\varphi$ is defined by setting $\chi_\varphi(g) = \mathrm{Tr}(\varphi_g)$. The character of an irreducible representation is called an *irreducible character*.

So if $\varphi \colon G \longrightarrow GL_n(\mathbb{C})$ is a representation given by $\varphi_g = (\varphi_{ij}(g))$, then

$$\chi_\varphi(g) = \sum_{i=1}^{n} \varphi_{ii}(g).$$

In general, to compute the character of a representation one must choose a basis and so when talking about characters, we may assume without loss of generality that we are talking about matrix representations.

*Remark 4.3.2.* If $\varphi \colon G \longrightarrow \mathbb{C}^*$ is a degree 1 representation, then $\chi_\varphi = \varphi$. From now on, we will not distinguish between a degree 1 representation and its character.

The first piece of information that we shall read off the character is the degree of the representation.

**Proposition 4.3.3.** *Let $\varphi$ be a representation of $G$. Then $\chi_\varphi(1) = \deg\varphi$.*

*Proof.* Indeed, suppose that $\varphi\colon G \longrightarrow GL(V)$ is a representation. Then $\mathrm{Tr}(\varphi_1) = \mathrm{Tr}(I) = \dim V = \deg\varphi$. □

A key property of the character is that it depends only on the equivalence class of the representation.

**Proposition 4.3.4.** *If $\varphi$ and $\rho$ are equivalent representations, then $\chi_\varphi = \chi_\rho$.*

*Proof.* Since the trace is computed by selecting a basis, we are able to assume that $\varphi, \rho\colon G \longrightarrow GL_n(\mathbb{C})$. Then, since they are equivalent, there is an invertible matrix $T \in GL_n(\mathbb{C})$ such that $\varphi_g = T\rho_g T^{-1}$, for all $g \in G$. Recalling that $\mathrm{Tr}(AB) = \mathrm{Tr}(BA)$, we obtain

$$\chi_\varphi(g) = \mathrm{Tr}(\varphi_g) = \mathrm{Tr}(T\rho_g T^{-1}) = \mathrm{Tr}(T^{-1}T\rho_g) = \mathrm{Tr}(\rho_g) = \chi_\rho(g)$$

as required. □

Essentially the same proof yields another crucial property of characters: they are constant on conjugacy classes.

**Proposition 4.3.5.** *Let $\varphi$ be a representation of $G$. Then, for all $g, h \in G$, the equality $\chi_\varphi(g) = \chi_\varphi(hgh^{-1})$ holds.*

*Proof.* Indeed, we compute

$$\begin{aligned}\chi_\varphi(hgh^{-1}) &= \mathrm{Tr}(\varphi_{hgh^{-1}}) = \mathrm{Tr}(\varphi_h\varphi_g\varphi_h^{-1})\\ &= \mathrm{Tr}(\varphi_h^{-1}\varphi_h\varphi_g) = \mathrm{Tr}(\varphi_g) = \chi_\varphi(g)\end{aligned}$$

again using $\mathrm{Tr}(AB) = \mathrm{Tr}(BA)$. □

Functions which are constant on conjugacy classes play an important role in representation theory and hence deserve a name of their own.

**Definition 4.3.6 (Class function).** A function $f\colon G \longrightarrow \mathbb{C}$ is called a *class function* if $f(g) = f(hgh^{-1})$ for all $g, h \in G$, or equivalently if $f$ is constant on conjugacy classes of $G$. The space of class functions is denoted $Z(L(G))$.

In particular, characters are class functions. The notation $Z(L(G))$ suggests that the class functions should be the center of some ring, and this will indeed be the case. If $f\colon G \longrightarrow \mathbb{C}$ is a class function and $C$ is a conjugacy class, $f(C)$ will denote the constant value that $f$ takes on $C$.

**Proposition 4.3.7.** *$Z(L(G))$ is a subspace of $L(G)$.*

*Proof.* Let $f_1, f_2$ be class functions on $G$ and let $c_1, c_2 \in \mathbb{C}$. Then

$$\begin{aligned}(c_1 f_1 + c_2 f_2)(hgh^{-1}) &= c_1 f_1(hgh^{-1}) + c_2 f_2(hgh^{-1}) \\ &= c_1 f_1(g) + c_2 f_2(g) = (c_1 f_1 + c_2 f_2)(g)\end{aligned}$$

showing that $c_1 f_1 + c_2 f_2$ is a class function. □

Next, let us compute the dimension of $Z(L(G))$. Let $Cl(G)$ be the set of conjugacy classes of $G$. Define, for $C \in Cl(G)$, the function $\delta_C \colon G \longrightarrow \mathbb{C}$ by

$$\delta_C(g) = \begin{cases} 1 & g \in C \\ 0 & g \notin C. \end{cases}$$

**Proposition 4.3.8.** *The set $B = \{\delta_C \mid C \in Cl(G)\}$ is a basis for $Z(L(G))$. Consequently,* $\dim Z(L(G)) = |Cl(G)|$.

*Proof.* Clearly, each $\delta_C$ is constant on conjugacy classes, and hence is a class function. Let us begin by showing that $B$ spans $Z(L(G))$. If $f \in Z(L(G))$, then one easily verifies that

$$f = \sum_{C \in Cl(G)} f(C)\delta_C.$$

Indeed, if $C'$ is the conjugacy class of $g$, then when you evaluate the right-hand side at $g$ you get $f(C')$. Since $g \in C'$, by definition $f(C') = f(g)$. To establish linear independence, we verify that $B$ is an orthogonal set of non-zero vectors. For if $C, C' \in Cl(G)$, then

$$\frac{1}{|G|}\sum_{g \in G} \delta_C(g)\overline{\delta_{C'}(g)} = \begin{cases} |C|/|G| & C = C' \\ 0 & C \neq C'. \end{cases}$$

This completes the proof that $B$ is a basis. Since $|B| = |Cl(G)|$, the calculation of the dimension follows. □

The next theorem is one of the fundamental results in group representation theory. It shows that the irreducible characters form an orthonormal set of class functions. This will be used to establish the uniqueness of the decomposition of a representation into irreducible constituents and to compute exactly the number of equivalence classes of irreducible representations.

**Theorem 4.3.9 (First orthogonality relations).** *Let $\varphi, \rho$ be irreducible representations of G. Then*

$$\langle \chi_\varphi, \chi_\rho \rangle = \begin{cases} 1 & \varphi \sim \rho \\ 0 & \varphi \nsim \rho. \end{cases}$$

*Thus the irreducible characters of G form an orthonormal set of class functions.*

*Proof.* Thanks to Propositions 3.2.4 and 4.3.4, we may assume without loss of generality that $\varphi\colon G \longrightarrow U_n(\mathbb{C})$ and $\rho\colon G \longrightarrow U_m(\mathbb{C})$ are unitary. Next we compute

$$\begin{aligned}\langle \chi_\varphi, \chi_\rho\rangle &= \frac{1}{|G|}\sum_{g\in G}\chi_\varphi(g)\overline{\chi_\rho(g)}\\ &= \frac{1}{|G|}\sum_{g\in G}\sum_{i=1}^{n}\varphi_{ii}(g)\sum_{j=1}^{m}\overline{\rho_{jj}}(G)\\ &= \sum_{i=1}^{n}\sum_{j=1}^{m}\frac{1}{|G|}\sum_{g\in G}\varphi_{ii}(g)\overline{\rho_{jj}(G)}\\ &= \sum_{i=1}^{n}\sum_{j=1}^{m}\langle \varphi_{ii}(g), \rho_{jj}(g)\rangle.\end{aligned}$$

The Schur orthogonality relations (Theorem 4.2.8) yield $\langle \varphi_{ii}(g), \rho_{jj}(g)\rangle = 0$ if $\varphi \nsim \rho$ and so $\langle \chi_\varphi, \chi_\rho\rangle = 0$ if $\varphi \nsim \rho$. If $\varphi \sim \rho$, then we may assume $\varphi = \rho$ by Proposition 4.3.4. In this case, the Schur orthogonality relations tell us

$$\langle \varphi_{ii}, \varphi_{jj}\rangle = \begin{cases} 1/n & i = j\\ 0 & i \neq j\end{cases}$$

and so

$$\langle \chi_\varphi, \chi_\varphi\rangle = \sum_{i=1}^{n}\langle \varphi_{ii}, \varphi_{ii}\rangle = \sum_{i=1}^{n}\frac{1}{n} = 1$$

as required. □

**Corollary 4.3.10.** *There are at most $|Cl(G)|$ equivalence classes of irreducible representations of G.*

*Proof.* First note that Theorem 4.3.9 implies inequivalent irreducible representations have distinct characters and, moreover, the irreducible characters form an orthonormal set. Since $\dim Z(L(G)) = |Cl(G)|$ and orthonormal sets are linearly independent, the corollary follows. □

Let us introduce some notation. If $V$ is a vector space, $\varphi$ is a representation and $m > 0$, then we set

$$mV = V \oplus \overbrace{\cdots}^{\times m} \oplus V \text{ and } m\varphi = \varphi \oplus \overbrace{\cdots}^{\times m} \oplus \varphi.$$

Let $\varphi^{(1)}, \ldots, \varphi^{(s)}$ be a complete set of irreducible unitary representations of $G$, up to equivalence. Again, set $d_i = \deg \varphi^{(i)}$.

**Definition 4.3.11 (Multiplicity).** If $\rho \sim m_1\varphi^{(1)} \oplus m_2\varphi^{(2)} \oplus \cdots \oplus m_s\varphi^{(s)}$, then $m_i$ is called the *multiplicity* of $\varphi^{(i)}$ in $\rho$. If $m_i > 0$, then we say that $\varphi^{(i)}$ is an *irreducible constituent* of $\rho$.

It is not clear at the moment that the multiplicity is well defined because we have not yet established the uniqueness of the decomposition of a representation into irreducibles. To show that it is well defined, we come up with a way to compute the multiplicity directly from the character of $\rho$. Since the character only depends on the equivalence class, it follows that the multiplicity of $\varphi^{(i)}$ will be the same no matter how we decompose $\rho$.

*Remark 4.3.12.* If $\rho \sim m_1\varphi^{(1)} \oplus m_2\varphi^{(2)} \oplus \cdots \oplus m_s\varphi^{(s)}$, then

$$\deg \rho = m_1 d_1 + m_2 d_2 + \cdots + m_s d_s$$

where we have retained the above notation.

**Lemma 4.3.13.** *Let* $\varphi = \rho \oplus \psi$. *Then* $\chi_\varphi = \chi_\rho + \chi_\psi$.

*Proof.* We may assume that $\rho\colon G \longrightarrow GL_n(\mathbb{C})$ and $\psi\colon G \longrightarrow GL_m(\mathbb{C})$. Then $\varphi\colon G \longrightarrow GL_{n+m}(\mathbb{C})$ has block form

$$\varphi_g = \begin{bmatrix} \rho_g & 0 \\ 0 & \psi_g \end{bmatrix}.$$

Since the trace is the sum of the diagonal elements, it follows that

$$\chi_\varphi(g) = \mathrm{Tr}(\varphi_g) = \mathrm{Tr}(\rho_g) + \mathrm{Tr}(\psi_g) = \chi_\rho(g) + \chi_\psi(g).$$

We conclude that $\chi_\varphi = \chi_\rho + \chi_\psi$. □

The above lemma implies that each character is an integral linear combination of irreducible characters. We can then use the orthonormality of the irreducible characters to extract the coefficients.

**Theorem 4.3.14.** *Let* $\varphi^{(1)}, \ldots, \varphi^{(s)}$ *be a complete set of representatives of the equivalence classes of irreducible representations of G and let*

$$\rho \sim m_1\varphi^{(1)} \oplus m_2\varphi^{(2)} \oplus \cdots \oplus m_s\varphi^{(s)}.$$

*Then* $m_i = \langle \chi_\rho, \chi_{\varphi^{(i)}} \rangle$. *Consequently, the decomposition of* $\rho$ *into irreducible constituents is unique and* $\rho$ *is determined up to equivalence by its character.*

*Proof.* By the previous lemma, $\chi_\rho = m_1\chi_{\varphi^{(1)}} + \cdots + m_s\chi_{\varphi^{(s)}}$. By the first orthogonality relations

$$\langle \chi_\rho, \chi_{\varphi^{(i)}} \rangle = m_1\langle \chi_{\varphi^{(1)}}, \chi_{\varphi^{(i)}} \rangle + \cdots + m_s\langle \chi_{\varphi^{(s)}}, \chi_{\varphi^{(i)}} \rangle = m_i,$$

proving the first statement. Proposition 4.3.4 implies the second and third statements. □

Theorem 4.3.14 offers a convenient criterion to check whether a representation is irreducible.

**Corollary 4.3.15.** *A representation $\rho$ is irreducible if and only if $\langle \chi_\rho, \chi_\rho \rangle = 1$.*

*Proof.* Suppose that $\rho \sim m_1\varphi^{(1)} \oplus m_2\varphi^{(2)} \oplus \cdots \oplus m_s\varphi^{(s)}$. Using the orthonormality of the irreducible characters, we obtain $\langle \chi_\rho, \chi_\rho \rangle = m_1^2 + \cdots + m_s^2$. The $m_i$ are non-negative integers, so $\langle \chi_\rho, \chi_\rho \rangle = 1$ if and only if there is an index $j$ so that $m_j = 1$ and $m_i = 0$ for $i \neq j$. But this happens precisely when $\rho$ is irreducible. □

Let us use Corollary 4.3.15 to show that the representation from Example 3.1.14 is irreducible.

*Example 4.3.16.* Let $\rho$ be the representation of $S_3$ from Example 3.1.14. Since $Id$, $(1\ 2)$ and $(1\ 2\ 3)$ form a complete set of representatives of the conjugacy classes of $S_3$, we can compute the inner product $\langle \chi_\rho, \chi_\rho \rangle$ from the values of the character on these elements. Now $\chi_\rho(Id) = 2$, $\chi_\rho((1\ 2)) = 0$ and $\chi_\rho((1\ 2\ 3)) = -1$. Since there are three transpositions and two 3-cycles, we have

$$\langle \chi_\rho, \chi_\rho \rangle = \frac{1}{6}\left(2^2 + 3 \cdot 0^2 + 2 \cdot (-1)^2\right) = 1.$$

Therefore, $\rho$ is irreducible.

Let us try to find all the irreducible characters of $S_3$ and to decompose the standard representation (cf. Example 3.1.9) into irreducibles.

*Example 4.3.17 (Characters of $S_3$).* We know that $S_3$ admits the trivial character $\chi_1 \colon S_3 \longrightarrow \mathbb{C}^*$ given by $\chi_1(\sigma) = 1$ for all $\sigma \in S_3$ (recall we identify a degree one representation with its character). We also have the character $\chi_3$ of the irreducible representation from Example 3.1.14. Since $S_3$ has three conjugacy classes, we might hope that there are three inequivalent irreducible representations of $S_3$. From Proposition 4.2.10, we know that if $d$ is the degree of the missing representation, then $1^2 + d^2 + 2^2 \leq 6$ and so $d = 1$. In fact, we can define a second degree one representation by

$$\chi_2(\sigma) = \begin{cases} 1 & \sigma \text{ is even} \\ -1 & \sigma \text{ is odd} \end{cases}$$

Let us form a table encoding this information (such a table is called a *character table*). The rows of Table 4.1 correspond to the irreducible characters, whereas the columns correspond to the conjugacy classes.

The standard representation of $S_3$ from Example 3.1.9 is given by the matrices

$$\varphi_{(1\ 2)} = \begin{bmatrix} 0 & 1 & 0 \\ 1 & 0 & 0 \\ 0 & 0 & 1 \end{bmatrix}, \quad \varphi_{(1\ 2\ 3)} = \begin{bmatrix} 0 & 0 & 1 \\ 1 & 0 & 0 \\ 0 & 1 & 0 \end{bmatrix}.$$

Hence we have character values as in Table 4.2.

**Table 4.1** Character table of $S_3$

| | $Id$ | (1 2) | (1 2 3) |
|---|---|---|---|
| $\chi_1$ | 1 | 1 | 1 |
| $\chi_2$ | 1 | $-1$ | 1 |
| $\chi_3$ | 2 | 0 | $-1$ |

**Table 4.2** The character $\chi_\varphi$

| | Id | (1 2) | (1 2 3) |
|---|---|---|---|
| $\chi_\varphi$ | 3 | 1 | 0 |

Inspection of Table 4.1 shows that $\chi_\varphi = \chi_1 + \chi_3$ and hence $\varphi \sim \chi_1 \oplus \rho$, as was advertised in Example 3.1.14. Alternatively, one could use Theorem 4.3.14 to obtain this result. Indeed,

$$\langle \chi_\varphi, \chi_1 \rangle = \frac{1}{6}(3 + 3 \cdot 1 + 2 \cdot 0) = 1$$
$$\langle \chi_\varphi, \chi_2 \rangle = \frac{1}{6}(3 + 3 \cdot (-1) + 2 \cdot 0) = 0$$
$$\langle \chi_\varphi, \chi_3 \rangle = \frac{1}{6}(6 + 3 \cdot 0 + 2 \cdot 0) = 1.$$

We will study the character table in detail later, in particular we shall show that the columns are always pairwise orthogonal, as is the case in Table 4.1.

## 4.4 The Regular Representation

Cayley's theorem asserts that $G$ is isomorphic to a subgroup of $S_n$ where $n = |G|$. The standard representation from Example 3.1.9 provides a representation $\varphi\colon S_n \longrightarrow GL_n(\mathbb{C})$. The restriction of this representation to $G$ will be called the regular representation of $G$, although we shall construct it formally speaking in a different way.

Let $X$ be a finite set. We build synthetically a vector space with basis $X$ by setting

$$\mathbb{C}X = \left\{ \sum_{x \in X} c_x x \mid c_x \in \mathbb{C} \right\}.$$

So $\mathbb{C}X$ consists of all formal linear combinations of elements of $X$. Two elements $\sum_{x \in X} a_x x$ and $\sum_{x \in X} b_x x$ are declared to be equal if and only if $a_x = b_x$ all $x \in X$. Addition is given by

$$\sum_{x \in X} a_x x + \sum_{x \in X} b_x x = \sum_{x \in X} (a_x + b_x) x;$$

scalar multiplication is defined similarly. We identify $x \in X$ with the linear combination $1 \cdot x$. Clearly $X$ is a basis for $\mathbb{C}X$. An inner product can be defined on $\mathbb{C}X$ by setting

$$\left\langle \sum_{x \in X} a_x x, \sum_{x \in X} b_x x \right\rangle = \sum_{x \in X} a_x \overline{b_x}.$$

**Definition 4.4.1 (Regular representation).** Let $G$ be a finite group. The *regular representation* of $G$ is the homomorphism $L\colon G \longrightarrow GL(\mathbb{C}G)$ defined by

$$L_g \sum_{h \in G} c_h h = \sum_{h \in G} c_h gh = \sum_{x \in G} c_{g^{-1}x} x, \tag{4.2}$$

for $g \in G$ (where the last equality comes from the change of variables $x = gh$).

The $L$ here stands for "left." Notice that on a basis element $h \in G$, we have $L_g h = gh$, i.e., $L_g$ acts on the basis via left multiplication by $g$. The formula in (4.2) is then the usual formula for a linear operator acting on a linear combination of basis vectors given the action on the basis. It follows that $L_g$ is a linear map for all $g \in G$. The regular representation is never irreducible when $G$ is non-trivial, but it has the positive feature that it contains all the irreducible representations of $G$ as constituents. Let us first prove that it is a representation.

**Proposition 4.4.2.** *The regular representation is a unitary representation of* $G$.

*Proof.* We have already pointed out that the map $L_g$ is linear for $g \in G$. Also if $g_1, g_2 \in G$ and $h \in G$ is a basis element of $\mathbb{C}G$, then

$$L_{g_1} L_{g_2} h = L_{g_1} g_2 h = g_1 g_2 h = L_{g_1 g_2} h$$

so that $L$ is a homomorphism. If we show that $L_g$ is unitary, it will then follow $L_g$ is invertible and that $L$ is a unitary representation. Now by (4.2)

$$\left\langle L_g \sum_{h \in G} c_h h, L_g \sum_{h \in G} k_h h \right\rangle = \left\langle \sum_{x \in G} c_{g^{-1}x} x, \sum_{x \in G} k_{g^{-1}x} x \right\rangle = \sum_{x \in G} c_{g^{-1}x} \overline{k_{g^{-1}x}}. \tag{4.3}$$

Setting $y = g^{-1}x$ turns the right-hand side of (4.3) into

$$\sum_{y \in G} c_y \overline{k_y} = \left\langle \sum_{y \in G} c_y y, \sum_{y \in G} k_y y \right\rangle$$

establishing that $L_g$ is unitary. □

Let us next compute the character of $L$. It turns out to have a particularly simple form.

**Proposition 4.4.3.** *The character of the regular representation $L$ is given by*

$$\chi_L(g) = \begin{cases} |G| & g = 1 \\ 0 & g \neq 1. \end{cases}$$

*Proof.* Let $G = \{g_1, \ldots, g_n\}$ where $n = |G|$. Then $L_g g_j = gg_j$. Thus if $[L_g]$ is the matrix of $L_g$ with respect to the basis $G$ with this ordering, then

$$\begin{aligned} [L_g]_{ij} &= \begin{cases} 1 & g_i = gg_j \\ 0 & \text{else} \end{cases} \\ &= \begin{cases} 1 & g = g_i g_j^{-1} \\ 0 & \text{else.} \end{cases} \end{aligned}$$

In particular,

$$[L_g]_{ii} = \begin{cases} 1 & g = 1 \\ 0 & \text{else} \end{cases}$$

from which we conclude

$$\chi_L(g) = \mathrm{Tr}(L_g) = \begin{cases} |G| & g = 1 \\ 0 & g \neq 1 \end{cases}$$

as required. □

We now decompose the regular representation $L$ into irreducible constituents. Fix again a complete set $\{\varphi^{(1)}, \ldots, \varphi^{(s)}\}$ of inequivalent irreducible unitary representations of our finite group $G$ and set $d_i = \deg \varphi^{(i)}$. For convenience, we put $\chi_i = \chi_{\varphi^{(i)}}$, for $i = 1, \ldots, s$.

**Theorem 4.4.4.** *Let $L$ be the regular representation of $G$. Then the decomposition*

$$L \sim d_1\varphi^{(1)} \oplus d_2\varphi^{(2)} \oplus \cdots \oplus d_s\varphi^{(s)}$$

*holds.*

*Proof.* We compute, using that $\chi_L(g) = 0$ for $g \neq 1$ and $\chi_L(1) = |G|$,

$$\begin{aligned} \langle \chi_L, \chi_i \rangle &= \frac{1}{|G|} \sum_{g \in G} \chi_L(g)\overline{\chi_i(g)} \\ &= \frac{1}{|G|} |G| \overline{\chi_i(1)} \\ &= \deg \varphi^{(i)} \\ &= d_i. \end{aligned}$$

This finishes the proof thanks to Theorem 4.3.14. □

With this theorem in hand, we can now complete the line of investigation initiated in this chapter.

**Corollary 4.4.5.** *The formula* $|G| = d_1^2 + d_2^2 + \cdots + d_s^2$ *holds.*

*Proof.* By Theorem 4.4.4, $\chi_L = d_1\chi_1 + d_2\chi_2 + \cdots + d_s\chi_s$. Thus evaluating $\chi_L$ at 1 yields $|G| = \chi_L(1) = d_1\chi_1(1) + \cdots + d_s\chi_s(1) = d_1^2 + \cdots + d_s^2$, as required. □

Consequently, we may infer that the matrix coefficients of irreducible unitary representations form an orthogonal basis for the space of all functions on $G$.

**Theorem 4.4.6.** *The set* $B = \{\sqrt{d_k}\varphi_{ij}^{(k)} \mid 1 \leq k \leq s, 1 \leq i, j \leq d_k\}$ *is an orthonormal basis for* $L(G)$, *where we have retained the above notation.*

*Proof.* We already know that $B$ is an orthonormal set by the orthogonality relations (Theorem 4.2.8). Since $|B| = d_1^2 + \cdots + d_s^2 = |G| = \dim L(G)$, it follows $B$ is a basis. □

Next we show that $\chi_1, \ldots, \chi_s$ is an orthonormal basis for the space of class functions.

**Theorem 4.4.7.** *The set* $\chi_1, \ldots, \chi_s$ *is an orthonormal basis for* $Z(L(G))$.

*Proof.* In this proof we retain the previous notation. The first orthogonality relations (Theorem 4.3.9) tell us that the irreducible characters form an orthonormal set of class functions. We must show that they span $Z(L(G))$. Let $f \in Z(L(G))$. By the previous theorem,

$$f = \sum_{i,j,k} c_{ij}^{(k)} \varphi_{ij}^{(k)}$$

for some $c_{ij}^{(k)} \in \mathbb{C}$ where $1 \leq k \leq s$ and $1 \leq i, j \leq d_k$. Since $f$ is a class function, for any $x \in G$, we have

$$\begin{aligned}
f(x) &= \frac{1}{|G|}\sum_{g\in G} f(g^{-1}xg) \\
&= \frac{1}{|G|}\sum_{g\in G}\sum_{i,j,k} c_{ij}^{(k)}\varphi_{ij}^{(k)}(g^{-1}xg) \\
&= \sum_{i,j,k} c_{ij}^{(k)} \frac{1}{|G|}\sum_{g\in G}\varphi_{ij}^{(k)}(g^{-1}xg) \\
&= \sum_{i,j,k} c_{ij}^{(k)} \left[\frac{1}{|G|}\sum_{g\in G}\varphi_{g^{-1}}^{(k)}\varphi_x^{(k)}\varphi_g^{(k)}\right]_{ij}
\end{aligned}$$

$$= \sum_{i,j,k} c_{ij}^{(k)} [(\varphi_x^{(k)})^\sharp]_{ij}$$

$$= \sum_{i,j,k} c_{ij}^{(k)} \frac{\mathrm{Tr}(\varphi_x^{(k)})}{\deg \varphi^{(k)}} I_{ij}$$

$$= \sum_{i,k} c_{ii}^{(k)} \frac{1}{d_k} \chi_k(x).$$

This establishes that

$$f = \sum_{i,k} c_{ii}^{(k)} \frac{1}{d_k} \chi_k$$

is in the span of $\chi_1, \ldots, \chi_s$, completing the proof that the irreducible characters form an orthonormal basis for $Z(L(G))$. □

**Corollary 4.4.8.** *The number of equivalence classes of irreducible representations of G is the number of conjugacy classes of G.*

*Proof.* The above theorem implies $s = \dim Z(L(G)) = |Cl(G)|$. □

**Corollary 4.4.9.** *A finite group G is abelian if and only if it has $|G|$ equivalence classes of irreducible representations.*

*Proof.* A finite group $G$ is abelian if and only if $|G| = |Cl(G)|$. □

*Example 4.4.10 (Irreducible representations of $\mathbb{Z}/n\mathbb{Z}$).* Let $\omega_n = e^{2\pi i/n}$. Define $\chi_k \colon \mathbb{Z}/n\mathbb{Z} \longrightarrow \mathbb{C}^*$ by $\chi_k([m]) = \omega_n^{km}$ for $0 \leq k \leq n-1$. Then $\chi_0, \ldots, \chi_{n-1}$ are the distinct irreducible representations of $\mathbb{Z}/n\mathbb{Z}$.

The representation theoretic information about a finite group $G$ can be encoded in a matrix known as its character table.

**Definition 4.4.11 (Character table).** Let $G$ be a finite group with irreducible characters $\chi_1, \ldots, \chi_s$ and conjugacy classes $C_1, \ldots, C_s$. The *character table* of $G$ is the $s \times s$ matrix $\mathsf{X}$ with $\mathsf{X}_{ij} = \chi_i(C_j)$. In other words, the rows of $\mathsf{X}$ are indexed by the characters of $G$, the columns by the conjugacy classes of $G$ and the $ij$-entry is the value of the $i$th-character on the $j$th-conjugacy class.

The character table of $S_3$ is recorded in Table 4.1, whereas that of $\mathbb{Z}/4\mathbb{Z}$ can be found in Table 4.3.

Notice that in both examples the columns are orthogonal with respect to the standard inner product. Let us prove that this is always the case. If $g, h \in G$, then the inner product of the columns corresponding to their conjugacy classes is given by the expression

$$\sum_{i=1}^{s} \chi_i(g)\overline{\chi_i(h)}.$$

**Table 4.3** Character table of $\mathbb{Z}/4\mathbb{Z}$

| | [0] | [1] | [2] | [3] |
|---|---|---|---|---|
| $\chi_1$ | 1 | 1 | 1 | 1 |
| $\chi_2$ | 1 | $-1$ | 1 | $-1$ |
| $\chi_3$ | 1 | $i$ | $-1$ | $-i$ |
| $\chi_4$ | 1 | $-i$ | $-1$ | $i$ |

Recall that if $C$ is a conjugacy class, then

$$\delta_C(g) = \begin{cases} 1 & g \in C \\ 0 & \text{else.} \end{cases}$$

The $\delta_C$ with $C \in Cl(G)$ form a basis for $Z(L(G))$, as do the irreducible characters. It is natural to express the $\delta_C$ in terms of the irreducible characters. This will yield the orthogonality of the columns of the character table.

**Theorem 4.4.12 (Second orthogonality relations).** *Let $C, C'$ be conjugacy classes of $G$ and let $g \in C$ and $h \in C'$. Then*

$$\sum_{i=1}^{s} \chi_i(g)\overline{\chi_i(h)} = \begin{cases} |G|/|C| & C = C' \\ 0 & C \neq C'. \end{cases}$$

*Consequently, the columns of the character table are orthogonal and hence the character table is invertible.*

*Proof.* Using that $\delta_{C'} = \sum_{i=1}^{s} \langle \delta_{C'}, \chi_i \rangle \chi_i$, we compute

$$\begin{aligned} \delta_{C'}(g) &= \sum_{i=1}^{s} \langle \delta_{C'}, \chi_i \rangle \chi_i(g) \\ &= \sum_{i=1}^{s} \frac{1}{|G|} \sum_{x \in G} \delta_{C'}(x) \overline{\chi_i(x)} \chi_i(g) \\ &= \sum_{i=1}^{s} \frac{1}{|G|} \sum_{x \in C'} \overline{\chi_i(x)} \chi_i(g) \\ &= \frac{|C'|}{|G|} \sum_{i=1}^{s} \chi_i(g) \overline{\chi_i(h)}. \end{aligned}$$

Since the left-hand side is 1 when $g \in C'$ and 0 otherwise, we conclude

$$\sum_{i=1}^{s} \chi_i(g)\overline{\chi_i(h)} = \begin{cases} |G|/|C| & C = C' \\ 0 & C \neq C' \end{cases}$$

as was required.

It now follows that the columns of the character table form an orthogonal set of non-zero vectors and hence are linearly independent. This yields the invertibility of the character table. □

*Remark 4.4.13.* The character table is in fact the transpose of the change of basis matrix from the basis $\{\chi_1, \ldots, \chi_s\}$ to the basis $\{\delta_C \mid C \in Cl(G)\}$ for $Z(L(G))$.

## 4.5 Representations of Abelian Groups

In this section, we compute the characters of an abelian group. Example 4.4.10 provides the characters of the group $\mathbb{Z}/n\mathbb{Z}$. Since any finite abelian group is a direct product of cyclic groups, all we need to know is how to compute the characters of a direct product of abelian groups. Let us proceed to the task at hand!

**Proposition 4.5.1.** *Let $G_1, G_2$ be abelian groups and suppose that $\chi_1, \ldots, \chi_m$ and $\varphi_1, \ldots, \varphi_n$ are the irreducible representations of $G_1, G_2$, respectively. In particular, $m = |G_1|$ and $n = |G_2|$. Then the functions $\alpha_{ij} \colon G_1 \times G_2 \longrightarrow \mathbb{C}^*$ with $1 \leq i \leq m$, $1 \leq j \leq n$ given by*

$$\alpha_{ij}(g_1, g_2) = \chi_i(g_1)\varphi_j(g_2)$$

*form a complete set of irreducible representations of $G_1 \times G_2$.*

*Proof.* First we check that the $\alpha_{ij}$ are homomorphisms. Indeed,

$$\begin{aligned}
\alpha_{ij}(g_1, g_2)\alpha_{ij}(g_1', g_2') &= \chi_i(g_1)\varphi_j(g_2)\chi_i(g_1')\varphi_j(g_2') \\
&= \chi_i(g_1)\chi_i(g_1')\varphi_j(g_2)\varphi_j(g_2') \\
&= \chi_i(g_1g_1')\varphi_j(g_2g_2') \\
&= \alpha_{ij}(g_1g_1', g_2g_2') \\
&= \alpha_{ij}((g_1, g_2)(g_1', g_2')).
\end{aligned}$$

Next we verify that $\alpha_{ij} = \alpha_{k\ell}$ implies $i = k$ and $j = \ell$. For if $\alpha_{ij} = \alpha_{k\ell}$, then

$$\chi_i(g) = \alpha_{ij}(g, 1) = \alpha_{k\ell}(g, 1) = \chi_k(g)$$

and so $i = k$. Similarly, $j = \ell$. As $G_1 \times G_2$ has $|G_1 \times G_2| = mn$ distinct irreducible representations, it follows that the $\alpha_{ij}$ with $1 \leq i \leq m$, $1 \leq j \leq n$ are all of them. □

*Example 4.5.2.* Let us compute the character table of the Klein four group $\mathbb{Z}/2\mathbb{Z} \times \mathbb{Z}/2\mathbb{Z}$. The character table of $\mathbb{Z}/2\mathbb{Z}$ is given in Table 4.4 and so using Proposition 4.5.1, the character table of $\mathbb{Z}/2\mathbb{Z} \times \mathbb{Z}/2\mathbb{Z}$ is as in Table 4.5.

Table 4.4 The character table of $\mathbb{Z}/2\mathbb{Z}$

| | [0] | [1] |
|---|---|---|
| $\chi_1$ | 1 | 1 |
| $\chi_2$ | 1 | −1 |

Table 4.5 The character table of $\mathbb{Z}/2\mathbb{Z} \times \mathbb{Z}/2\mathbb{Z}$

| | ([0], [0]) | ([0], [1]) | ([1], [0]) | ([1], [1]) |
|---|---|---|---|---|
| $\alpha_{11}$ | 1 | 1 | 1 | 1 |
| $\alpha_{12}$ | 1 | −1 | 1 | −1 |
| $\alpha_{21}$ | 1 | 1 | −1 | −1 |
| $\alpha_{22}$ | 1 | −1 | −1 | 1 |

# Exercises

**Exercise 4.1.** Let $\varphi\colon G \longrightarrow GL(U)$, $\psi\colon G \longrightarrow GL(V)$ and $\rho\colon G \longrightarrow GL(W)$ be representations of a group $G$. Suppose that $T \in \mathrm{Hom}_G(\varphi, \psi)$ and $S \in \mathrm{Hom}_G(\psi, \rho)$. Prove that $ST \in \mathrm{Hom}_G(\varphi, \rho)$.

**Exercise 4.2.** Let $\varphi$ be a representation of a group $G$ with character $\chi_\varphi$. Prove that $\chi_\varphi(g^{-1}) = \overline{\chi_\varphi(g)}$.

**Exercise 4.3.** Let $\varphi\colon G \longrightarrow GL(V)$ be an irreducible representation. Let

$$Z(G) = \{a \in G \mid ag = ga, \forall g \in G\}$$

be the center of $G$. Show that if $a \in Z(G)$, then $\varphi(a) = \lambda I$ for some $\lambda \in \mathbb{C}^*$.

**Exercise 4.4.** Let $G$ be a group. Show that $f\colon G \longrightarrow \mathbb{C}$ is a class function if and only if $f(gh) = f(hg)$ for all $g, h \in G$.

**Exercise 4.5.** For $v = (c_1, \ldots, c_m) \in (\mathbb{Z}/2\mathbb{Z})^m$, let $\alpha(v) = \{i \mid c_i = [1]\}$. For each $Y \subseteq \{1, \ldots, m\}$, define a function $\chi_Y\colon (\mathbb{Z}/2\mathbb{Z})^m \longrightarrow \mathbb{C}$ by

$$\chi_Y(v) = (-1)^{|\alpha(v) \cap Y|}.$$

1. Show that $\chi_Y$ is a character.
2. Show that every irreducible character of $(\mathbb{Z}/2\mathbb{Z})^m$ is of the form $\chi_Y$ for some subset $Y$ of $\{1, \ldots, m\}$.
3. Show that if $X, Y \subseteq \{1, \ldots, m\}$, then $\chi_X(v)\chi_Y(v) = \chi_{X \triangle Y}(v)$ where $X \triangle Y = X \cup Y \setminus (X \cap Y)$ is the symmetric difference.

**Exercise 4.6.** Let $\mathrm{sgn}\colon S_n \longrightarrow \mathbb{C}^*$ be the representation given by

$$\mathrm{sgn}(\sigma) = \begin{cases} 1 & \sigma \text{ is even} \\ -1 & \sigma \text{ is odd.} \end{cases}$$

Show that if $\chi$ is the character of an irreducible representation of $S_n$ not equivalent to sgn, then

$$\sum_{\sigma \in S_n} \operatorname{sgn}(\sigma)\chi(\sigma) = 0.$$

**Exercise 4.7.** Let $\varphi\colon G \longrightarrow GL_n(\mathbb{C})$ and $\rho\colon G \longrightarrow GL_m(\mathbb{C})$ be representations. Let $V = M_{mn}(\mathbb{C})$. Define $\tau\colon G \longrightarrow GL(V)$ by $\tau_g(A) = \rho_g A \varphi_g^T$.

1. Show that $\tau$ is a representation of $G$.
2. Show that
$$\tau_g E_{k\ell} = \sum_{i,j} \rho_{ik}(g)\varphi_{j\ell}(g)E_{ij}.$$
3. Prove that $\chi_\tau(g) = \chi_\rho(g)\chi_\varphi(g)$.
4. Conclude that the pointwise product of two characters of $G$ is a character of $G$.

**Exercise 4.8.** Let $\alpha\colon S_n \longrightarrow GL_n(\mathbb{C})$ be the standard representation from Example 3.1.9.

1. Show that $\chi_\alpha(\sigma)$ is the number of fixed points of $\sigma$, that is, the number of elements $k \in \{1, \ldots, n\}$ such that $\sigma(k) = k$.
2. Show that if $n = 3$, then $\langle \chi_\alpha, \chi_\alpha \rangle = 2$ and hence $\alpha$ is not irreducible.

**Exercise 4.9.** Let $\chi$ be a non-trivial irreducible character of a finite group $G$. Show that

$$\sum_{g \in G} \chi(g) = 0.$$

**Exercise 4.10.** Let $\varphi\colon G \longrightarrow H$ be a surjective homomorphism and let $\psi\colon H \longrightarrow GL(V)$ be an irreducible representation. Prove that $\psi \circ \varphi$ is an irreducible representation of $G$.

**Exercise 4.11.** Let $G_1$ and $G_2$ be finite groups and let $G = G_1 \times G_2$. Suppose $\rho\colon G_1 \longrightarrow GL_m(\mathbb{C})$ and $\varphi\colon G_2 \longrightarrow GL_n(\mathbb{C})$ are representations. Let $V = M_{mn}(\mathbb{C})$. Define $\tau\colon G \longrightarrow GL(V)$ by $\tau_{(g_1,g_2)}(A) = \rho_{g_1} A \varphi_{g_2}^T$.

1. Show that $\tau$ is a representation of $G$.
2. Prove that $\chi_\tau(g_1, g_2) = \chi_\rho(g_1)\chi_\varphi(g_2)$.
3. Show that if $\rho$ and $\varphi$ are irreducible, then $\tau$ is irreducible.
4. Prove that every irreducible representation of $G_1 \times G_2$ can be obtained in this way.

**Exercise 4.12.** Let $Q = \{\pm 1, \pm \hat{\imath}, \pm \hat{\jmath}, \pm \hat{k}\}$ be the group of quaternions. The key rules to know are that $\hat{\imath}^2 = \hat{\jmath}^2 = \hat{k}^2 = \hat{\imath}\hat{\jmath}\hat{k} = -1$.

1. Show that $\rho\colon Q \longrightarrow GL_2(\mathbb{C})$ defined by

$$\rho(\pm 1) = \pm \begin{bmatrix} 1 & 0 \\ 0 & 1 \end{bmatrix}, \quad \rho(\pm \hat{\imath}) = \pm \begin{bmatrix} i & 0 \\ 0 & -i \end{bmatrix},$$

$$\rho(\pm \hat{\jmath}) = \pm \begin{bmatrix} 0 & 1 \\ -1 & 0 \end{bmatrix}, \quad \rho(\pm \hat{k}) = \begin{bmatrix} 0 & i \\ i & 0 \end{bmatrix}$$

   is an irreducible representation of $Q$.
2. Find four inequivalent degree one representations of $Q$. (Hint: $N = \{\pm 1\}$ is a normal subgroup of $Q$ and $Q/N \cong \mathbb{Z}/2\mathbb{Z} \times \mathbb{Z}/2\mathbb{Z}$. Use this to obtain the four inequivalent representations of degree 1.)
3. Show that the conjugacy classes of $Q$ are $\{1\}$, $\{-1\}$, $\{\pm \hat{\imath}\}$, $\{\pm \hat{\jmath}\}$, and $\{\pm \hat{k}\}$.
4. Write down the character table for $Q$.

**Exercise 4.13.** Let $G$ be a group and let $G'$ be the commutator subgroup of $G$. That is, $G'$ is the subgroup of $G$ generated by all commutators $[g, h] = g^{-1}h^{-1}gh$ with $g, h \in G$. You may take for granted the following facts that are typically proved in a first course in group theory:

1. $G'$ is a normal subgroup of $G$.
2. $G/G'$ is an abelian group.
3. If $N$ is a normal subgroup of $G$, then $G/N$ is an abelian if and only if $G' \subseteq N$.

Let $\varphi\colon G \longrightarrow G/G'$ be the canonical homomorphism given by $\varphi(g) = gG'$. Prove that every degree one representation $\rho\colon G \longrightarrow \mathbb{C}^*$ is of the form $\psi \circ \varphi$ where $\psi\colon G/G' \longrightarrow \mathbb{C}^*$ is a degree one representation of the abelian group $G/G'$.

**Exercise 4.14.** Show that if $G$ is a finite group and $g$ is a non-trivial element of $G$, then there is an irreducible representation $\varphi$ with $\varphi(g) \neq I$. (Hint: Let $L\colon G \longrightarrow GL(\mathbb{C}G)$ be the regular representation. Show that $L_g \neq I$. Use the decomposition of $L$ into irreducible representations to show that $\varphi_g \neq I$ for some irreducible.)

**Exercise 4.15.** This exercise provides an alternate proof of the orthogonality relations for characters. It requires Exercise 3.5. Let $\varphi\colon G \longrightarrow GL_m(\mathbb{C})$ and $\psi\colon G \longrightarrow GL_n(\mathbb{C})$ be representations of a finite group $G$. Let $V = M_{mn}(\mathbb{C})$ and define a representation $\rho\colon G \longrightarrow GL(V)$ by $\rho_g(A) = \varphi_g A \psi_g^*$.

1. Prove that $\chi_\rho(g) = \chi_\varphi(g)\overline{\chi_\psi(g)}$ for $g \in G$.
2. Show that $V^G = \mathrm{Hom}_G(\psi, \varphi)$ (where $V^G$ is defined as per Exercise 3.5).
3. Deduce $\dim \mathrm{Hom}_G(\psi, \varphi) = \langle \chi_\varphi, \chi_\psi \rangle$ using Exercise 3.5.
4. Deduce using Schur's lemma that if $\varphi$ and $\psi$ are irreducible representations, then

$$\langle \chi_\varphi, \chi_\psi \rangle = \begin{cases} 1 & \varphi \sim \psi \\ 0 & \varphi \nsim \psi. \end{cases}$$

# Chapter 5
# Fourier Analysis on Finite Groups

In this chapter we introduce an algebraic structure on $L(G)$ coming from the convolution product. The Fourier transform then permits us to analyze this structure more clearly in terms of known rings. In particular, we prove Wedderburn's theorem for group algebras over the complex numbers. Due to its applications in signal and image processing, statistics [3, 7, 8, 22], combinatorics, and number theory, Fourier analysis is one of the most important aspects of mathematics. There are entire books dedicated to Fourier analysis on finite groups [3, 22]. Unfortunately, we merely scratch the surface of this rich theory in this text. In this chapter, the only application that we give is to the computation of the eigenvalues of the adjacency matrix of a Cayley graph of an abelian group. In Chap. 11, we present some applications to probability theory.

## 5.1 Periodic Functions on Cyclic Groups

We begin with the classical case of periodic functions on the integers.

**Definition 5.1.1 (Periodic function).** A function $f\colon \mathbb{Z} \longrightarrow \mathbb{C}$ is said to be *periodic* with period $n$ if $f(x) = f(x+n)$ for all $x \in \mathbb{Z}$.

Notice that if $n$ is a period for $f$, then so is any multiple of $n$. It is easy to see that periodic functions with period $n$ are in bijection with elements of $L(\mathbb{Z}/n\mathbb{Z})$, that is, functions $f\colon \mathbb{Z}/n\mathbb{Z} \longrightarrow \mathbb{C}$. Indeed, the definition of a periodic function says precisely that $f$ is constant on residue classes modulo $n$. Now the irreducible characters form a basis for $L(\mathbb{Z}/n\mathbb{Z})$ and are given in Example 4.4.10. It follows that if $f\colon \mathbb{Z}/n\mathbb{Z} \longrightarrow \mathbb{C}$ is a function, then

$$f = \langle f, \chi_0 \rangle \chi_0 + \cdots + \langle f, \chi_{n-1} \rangle \chi_{n-1} \tag{5.1}$$

where $\chi_k([m]) = e^{2\pi i k m/n}$. The Fourier transform encodes this information as a function.

B. Steinberg, *Representation Theory of Finite Groups: An Introductory Approach*, Universitext, DOI 10.1007/978-1-4614-0776-8_5,
© Springer Science+Business Media, LLC 2012

**Definition 5.1.2 (Fourier transform).** Let $f\colon \mathbb{Z}/n\mathbb{Z} \longrightarrow \mathbb{C}$. Define the *Fourier transform* $\widehat{f}\colon \mathbb{Z}/n\mathbb{Z} \longrightarrow \mathbb{C}$ of $f$ by

$$\widehat{f}([m]) = n\langle f, \chi_m\rangle = \sum_{k=0}^{n-1} f([k])e^{-2\pi imk/n}$$

It is immediate that the Fourier transform $T\colon L(\mathbb{Z}/n\mathbb{Z}) \longrightarrow L(\mathbb{Z}/n\mathbb{Z})$ is linear because of the linearity of inner products in the first variable. We can rewrite (5.1) as:

**Proposition 5.1.3 (Fourier inversion).** *The Fourier transform is invertible. More precisely,* $f = \frac{1}{n}\sum_{k=0}^{n-1}\widehat{f}([k])\chi_k$.

The Fourier transform on cyclic groups is used in signal and image processing. The idea is that the values of $\widehat{f}$ correspond to the wavelengths associated to the wave function $f$. One sets to zero all sufficiently small values of $\widehat{f}$, thereby compressing the wave. To recover something close enough to the original wave, as far as our eyes and ears are concerned, one applies Fourier inversion.

## 5.2 The Convolution Product

We now introduce the convolution product on $L(G)$, thereby explaining the terminology group algebra for $L(G)$.

**Definition 5.2.1 (Convolution).** Let $G$ be a finite group and $a, b \in L(G)$. Then the *convolution* $a * b\colon G \longrightarrow \mathbb{C}$ is defined by

$$a * b(x) = \sum_{y\in G} a(xy^{-1})b(y). \tag{5.2}$$

Our eventual goal is to show that convolution gives $L(G)$ the structure of a ring. Before that, let us motivate the definition of convolution. To each element $g \in G$, we have associated the delta function $\delta_g$. What could be more natural than to try and assign a multiplication $*$ to $L(G)$ so that $\delta_g * \delta_h = \delta_{gh}$? Let us show that convolution has this property. Indeed,

$$\delta_g * \delta_h(x) = \sum_{y\in G} \delta_g(xy^{-1})\delta_h(y)$$

and the only non-zero term is when $y = h$ and $g = xy^{-1} = xh^{-1}$, i.e., $x = gh$. In this case, one gets 1, so we have proved:

**Proposition 5.2.2.** *For* $g, h \in G$, $\delta_g * \delta_h = \delta_{gh}$. □

Now if $a, b \in L(G)$, then

$$a = \sum_{g \in G} a(g)\delta_g, \; b = \sum_{g \in G} b(g)\delta_g$$

so if $L(G)$ were really a ring, then the distributive law would yield

$$a * b = \sum_{g,h \in G} a(g)b(h)\delta_g * \delta_h = \sum_{g,h \in G} a(g)b(h)\delta_{gh}.$$

Applying the change of variables $x = gh$, $y = h$ then gives us

$$a * b = \sum_{x \in G} \left( \sum_{y \in G} a(xy^{-1})b(y) \right) \delta_x$$

which is equivalent to the formula (5.2). Another motivation for the definition of convolution comes from statistics; see Chap. 11.

**Theorem 5.2.3.** *The set $L(G)$ is a ring with addition taken pointwise and convolution as multiplication. Moreover, $\delta_1$ is the multiplicative identity.*

*Proof.* We will only verify that $\delta_1$ is the identity and the associativity of convolution. The remaining verifications that $L(G)$ is a ring are straightforward and will be left to the reader. Let $a \in L(G)$. Then

$$a * \delta_1(x) = \sum_{y \in G} a(xy^{-1})\delta_1(y^{-1}) = a(x)$$

since $\delta_1(y^{-1}) = 0$ except when $y = 1$. Similarly, $\delta_1 * a = a$. This proves $\delta_1$ is the identity.

For associativity, let $a, b, c \in L(G)$. Then

$$[(a * b) * c](x) = \sum_{y \in G} [a * b](xy^{-1})c(y) = \sum_{y \in G} \sum_{z \in G} a(xy^{-1}z^{-1})b(z)c(y). \quad (5.3)$$

We make the change of variables $u = zy$ (and so $y^{-1}z^{-1} = u^{-1}$, $z = uy^{-1}$). The right-hand side of (5.3) then becomes

$$\begin{aligned} \sum_{y \in G} \sum_{u \in G} a(xu^{-1})b(uy^{-1})c(y) &= \sum_{u \in G} a(xu^{-1}) \sum_{y \in G} b(uy^{-1})c(y) \\ &= \sum_{u \in G} a(xu^{-1})[b * c](u) \\ &= [a * (b * c)](x) \end{aligned}$$

completing the proof. □

It is now high time to justify the notation $Z(L(G))$ for the space of class functions on $G$. Recall that the center $Z(R)$ of a ring $R$ consists of all elements $a \in R$ such that $ab = ba$ for all $b \in R$. For instance, one can show that the scalar matrices form the center of $M_n(\mathbb{C})$.

**Proposition 5.2.4.** *The class functions form the center of $L(G)$. That is, $f\colon G \longrightarrow \mathbb{C}$ is a class function if and only if $a * f = f * a$ for all $a \in L(G)$.*

*Proof.* Suppose first that $f$ is a class function and let $a \in L(G)$. Then

$$a * f(x) = \sum_{y \in G} a(xy^{-1}) f(y) = \sum_{y \in G} a(xy^{-1}) f(xyx^{-1}) \tag{5.4}$$

since $f$ is a class function. Setting $z = xy^{-1}$ turns the right-hand side of (5.4) into

$$\sum_{z \in G} a(z) f(xz^{-1}) = \sum_{z \in G} f(xz^{-1}) a(z) = f * a(x)$$

and hence $a * f = f * a$.

For the other direction, let $f$ be in the center of $L(G)$.

*Claim.* $f(gh) = f(hg)$ for all $g, h \in G$.

*Proof (of claim).* Observe that

$$\begin{aligned} f(gh) &= \sum_{y \in G} f(gy^{-1}) \delta_{h^{-1}}(y) = f * \delta_{h^{-1}}(g) \\ &= \delta_{h^{-1}} * f(g) = \sum_{y \in G} \delta_{h^{-1}}(gy^{-1}) f(y) = f(hg) \end{aligned}$$

since $\delta_{h^{-1}}(gy^{-1})$ is non-zero if and only if $gy^{-1} = h^{-1}$, that is, $y = hg$. □

To complete the proof, we note that by the claim $f(ghg^{-1}) = f(hg^{-1}g) = f(h)$, establishing that $f$ is a class function. □

As a consequence of this result, our notation $Z(L(G))$ for the set of class functions is unambiguous.

## 5.3 Fourier Analysis on Finite Abelian Groups

In this section, we consider the case of abelian groups as the situation is much simpler, and is frequently sufficient for applications to signal processing and number theory. In number theory, the groups of interest are usually $\mathbb{Z}/n\mathbb{Z}$ and $\mathbb{Z}/n\mathbb{Z}^*$.

Let $G$ be a finite abelian group. Then class functions on $G$ are the same thing as functions, that is, $L(G) = Z(L(G))$. Therefore, $L(G)$ is a commutative ring. Let us try to identify it (up to isomorphism) with a known ring. The secret to analyzing the ring structure on $L(G)$ is the Fourier transform.

**Definition 5.3.1 (Dual group).** Let $G$ be a finite abelian group and let $\widehat{G}$ be the set of all irreducible characters $\chi\colon G \longrightarrow \mathbb{C}^*$. One calls $\widehat{G}$ the *dual group* of $G$.

Of course, given the name dual group, we should prove that it is a group, although we will not use this fact.

**Proposition 5.3.2.** *Let $G$ be a finite abelian group. Define a product on $\widehat{G}$ via pointwise multiplication, that is, $(\chi \cdot \theta)(g) = \chi(g)\theta(g)$. Then $\widehat{G}$ is an abelian group of order $|G|$ with respect to this binary operation.*

*Proof.* First observe that if $\chi, \theta \in \widehat{G}$, then

$$\begin{aligned}\chi \cdot \theta(g_1 g_2) &= \chi(g_1 g_2)\theta(g_1 g_2) = \chi(g_1)\chi(g_2)\theta(g_1)\theta(g_2)\\ &= \chi(g_1)\theta(g_1)\chi(g_2)\theta(g_2) = (\chi \cdot \theta)(g_1) \cdot (\chi \cdot \theta)(g_2)\end{aligned}$$

and so $\widehat{G}$ is closed under the pointwise product. Trivially, the product is associative and commutative. The identity is the trivial character $\chi_1(g) = 1$ for all $g \in G$. The inverse is given by $\chi^{-1}(g) = \chi(g)^{-1} = \overline{\chi(g)}$ (as $\chi$ is unitary). One easily verifies that $\chi^{-1}$ is a character. Trivially, $\chi \cdot \chi^{-1} = \chi_1$. Thus $\widehat{G}$ is an abelian group. We already know that the number of irreducible characters of $G$ is $|G|$. This completes the proof. □

*Example 5.3.3.* Let $G = \mathbb{Z}/n\mathbb{Z}$. Then $\widehat{G} = \{\chi_0, \ldots, \chi_{n-1}\}$ where

$$\chi_k([m]) = e^{2\pi i k m/n}.$$

One easily verifies that the assignment $[k] \mapsto \chi_k$ is a group isomorphism $G \longrightarrow \widehat{G}$.

The above example is typical: one always has that $G \cong \widehat{G}$. This can be deduced from Example 5.3.3, the results of Sect. 4.5 and the fact that each finite abelian group is a direct product of cyclic groups. We leave a proof of this to the exercises.

We now introduce a vector space isomorphism $L(G) \longrightarrow L(\widehat{G})$ called the Fourier transform. To make it a ring homomorphism, we have to use a different product on $L(\widehat{G})$ than convolution.

**Definition 5.3.4 (Fourier transform).** Let $f\colon G \longrightarrow \mathbb{C}$ be a complex-valued function on a finite abelian group $G$. Then the *Fourier transform* $\widehat{f}\colon \widehat{G} \longrightarrow \mathbb{C}$ is defined by

$$\widehat{f}(\chi) = |G|\langle f, \chi\rangle = \sum_{g\in G} f(g)\overline{\chi(g)}.$$

The complex numbers $|G|\langle f, \chi\rangle$ are often called the *Fourier coefficients* of $f$.

In Sect. 5.1, we defined the Fourier transform in a different way for cyclic groups. However, it amounts to the same thing under the isomorphism between $\widehat{G}$ and $G$ considered in Example 5.3.3.

*Example 5.3.5.* If $\chi, \theta \in \widehat{G}$, then

$$\widehat{\chi}(\theta) = |G|\langle \chi, \theta \rangle = \begin{cases} |G| & \chi = \theta \\ 0 & \text{else} \end{cases}$$

by the orthogonality relations and so $\widehat{\chi} = |G|\delta_{\chi}$.

**Theorem 5.3.6 (Fourier inversion).** *If $f \in L(G)$, then*

$$f = \frac{1}{|G|} \sum_{\chi \in \widehat{G}} \widehat{f}(\chi)\chi.$$

*Proof.* The proof is a straightforward computation:

$$f = \sum_{\chi \in \widehat{G}} \langle f, \chi \rangle \chi = \frac{1}{|G|} \sum_{\chi \in \widehat{G}} |G|\langle f, \chi \rangle \chi = \frac{1}{|G|} \sum_{\chi \in \widehat{G}} \widehat{f}(\chi)\chi$$

as required. □

Next we observe that the Fourier transform is a linear mapping.

**Proposition 5.3.7.** *The map $T \colon L(G) \longrightarrow L(\widehat{G})$ given by $Tf = \widehat{f}$ is an invertible linear transformation.*

*Proof.* Let $|G| = n$. By definition $T(c_1 f_1 + c_2 f_2) = \widehat{c_1 f_1 + c_2 f_2}$. Now

$$\begin{aligned} \widehat{c_1 f_1 + c_2 f_2}(\chi) &= n\langle c_1 f_1 + c_2 f_2, \chi \rangle \\ &= c_1 n\langle f_1, \chi \rangle + c_2 n\langle f_2, \chi \rangle \\ &= c_1 \widehat{f_1}(\chi) + c_2 \widehat{f_2}(\chi) \end{aligned}$$

and so $\widehat{c_1 f_1 + c_2 f_2} = c_1 \widehat{f_1} + c_2 \widehat{f_2}$, establishing that $T$ is linear. Theorem 5.3.6 implies $T$ is injective and hence invertible since $\dim L(G) = n = \dim L(\widehat{G})$. □

Let $A$ be an abelian group. There are two ways to make $L(A)$ into a ring: one way is via convolution; the other is to use pointwise multiplication: $(f \cdot g)(x) = f(x)g(x)$. The reader should observe that $\delta_1$ is the identity for convolution and that the constant map to 1 is the identity for the pointwise product. The next theorem shows that the Fourier transform gives an isomorphism between these two ring structures, that is, it sends convolution to pointwise multiplication.

**Theorem 5.3.8.** *The Fourier transform satisfies*

$$\widehat{a * b} = \widehat{a} \cdot \widehat{b}.$$

*Consequently, the linear map $T\colon L(G) \longrightarrow L(\widehat{G})$ given by $Tf = \widehat{f}$ provides a ring isomorphism between $(L(G), +, *)$ and $(L(\widehat{G}), +, \cdot)$.*

*Proof.* We know by Proposition 5.3.7 that $T$ is an isomorphism of vector spaces. Therefore, to show that it is a ring isomorphism it suffices to show $T(a*b) = Ta \cdot Tb$, that is, $\widehat{a * b} = \widehat{a} \cdot \widehat{b}$. Let us proceed to do this. Set $n = |G|$.

$$
\begin{aligned}
\widehat{a * b}(\chi) &= n\langle a * b, \chi\rangle \\
&= n \cdot \frac{1}{n} \sum_{x \in G} (a * b)(x)\overline{\chi(x)} \\
&= \sum_{x \in G} \overline{\chi(x)} \sum_{y \in G} a(xy^{-1})b(y) \\
&= \sum_{y \in G} b(y) \sum_{x \in G} a(xy^{-1})\overline{\chi(x)}.
\end{aligned}
$$

Changing variables, we put $z = xy^{-1}$ (and so $x = zy$). Then we obtain

$$
\begin{aligned}
\widehat{a * b}(\chi) &= \sum_{y \in G} b(y) \sum_{z \in G} a(z)\overline{\chi(zy)} \\
&= \sum_{y \in G} b(y)\overline{\chi(y)} \sum_{z \in G} a(z)\overline{\chi(z)} \\
&= \sum_{z \in G} a(z)\overline{\chi(z)} \sum_{y \in G} b(y)\overline{\chi(y)} \\
&= n\langle a, \chi\rangle \cdot n\langle b, \chi\rangle \\
&= \widehat{a}(\chi)\widehat{b}(\chi)
\end{aligned}
$$

and so $\widehat{a * b} = \widehat{a} \cdot \widehat{b}$, as was required. □

Let us summarize what we have proved for the classical case of periodic functions on $\mathbb{Z}$, where we identify $\mathbb{Z}/n\mathbb{Z}$ with $\widehat{\mathbb{Z}/n\mathbb{Z}}$ via $[k] \mapsto \chi_k$.

*Example 5.3.9 (Periodic functions on $\mathbb{Z}$).* Let $f, g\colon \mathbb{Z} \longrightarrow \mathbb{C}$ have period $n$. Their convolution is defined by

$$
f * g(m) = \sum_{k=0}^{n-1} f(m-k)g(k).
$$

The Fourier transform is then

$$
\widehat{f}(m) = \sum_{k=0}^{n-1} f(k)e^{-2\pi i mk/n}.
$$

The Fourier inversion theorem says that

$$f(m) = \frac{1}{n}\sum_{k=0}^{n-1} \widehat{f}(k)e^{2\pi imk/n}.$$

The multiplication formula says that $\widehat{f*g} = \widehat{f}\cdot\widehat{g}$. In practice it is more efficient to compute $\widehat{f}\cdot\widehat{g}$ and then apply Fourier inversion to obtain $f*g$ than to compute $f*g$ directly thanks to the existence of the fast Fourier transform.

The original Fourier transform was invented by Fourier in the continuous context in the early 1800s to study the heat equation. For absolutely integrable complex-valued functions $f, g\colon \mathbb{R} \longrightarrow \mathbb{C}$, their convolution is defined by

$$f*g(x) = \int_{-\infty}^{\infty} f(x-y)g(y)dy.$$

The Fourier transform of $f$ is

$$\widehat{f}(x) = \int_{-\infty}^{\infty} f(t)e^{-2\pi ixt}dt.$$

Fourier inversion says that

$$f(x) = \int_{-\infty}^{\infty} \widehat{f}(t)e^{2\pi ixt}dt.$$

Once again the multiplication rule $\widehat{f*g} = \widehat{f}\cdot\widehat{g}$ holds.

## 5.4 An Application to Graph Theory

A *graph* $\varGamma$ consists of a set $V$ of *vertices* and a set $E$ of unordered pairs of elements of $V$, called *edges*. We shall only consider finite graphs in this section. One often views graphs pictorially by representing each vertex as a point and drawing a line segment between two vertices that form an edge.

For instance, if $\varGamma$ has vertex set $V=\{1,2,3,4\}$ and edge set $E=\{\{1,3\},\{2,3\},\{2,4\},\{3,4\}\}$, then the picture is as in Fig. 5.1.

One can usefully encode a graph by its adjacency matrix.

**Definition 5.4.1 (Adjacency matrix).** Let $\varGamma$ be a graph with vertex set $V = \{v_1,\ldots,v_n\}$ and edge set $E$. Then the adjacency matrix $A = (a_{ij})$ is given by

$$a_{ij} = \begin{cases} 1 & \{v_i,v_j\} \in E \\ 0 & \text{else.} \end{cases}$$

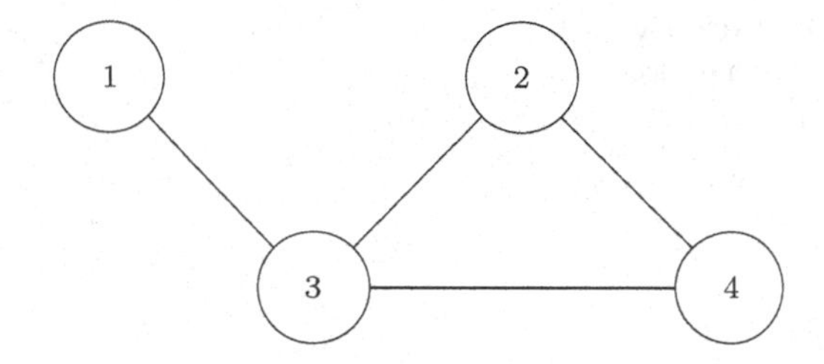

**Fig. 5.1** A graph

*Example 5.4.2.* For the graph in Fig. 5.1, the adjacency matrix is

$$A = \begin{bmatrix} 0 & 0 & 1 & 0 \\ 0 & 0 & 1 & 1 \\ 1 & 1 & 0 & 1 \\ 0 & 1 & 1 & 0 \end{bmatrix}.$$

Notice that the adjacency matrix is always symmetric and hence diagonalizable with real eigenvalues by the spectral theorem for matrices. The set of eigenvalues of $A$ is called the *spectrum* of the graph; it does not depend on the ordering of the vertices. One can obtain important information from the eigenvalues, such as the number of spanning trees. Also, one can verify that $A^n_{ij}$ is the number of paths of length $n$ from $v_i$ to $v_j$. For a diagonalizable matrix, knowing the eigenvalues already gives a lot of information about powers of the matrix. There is a whole area of graph theory, called spectral graph theory, dedicated to studying graphs via their eigenvalues. The adjacency matrix is also closely related to the study of random walks on the graph.

A natural source of graphs, known as Cayley graphs, comes from group theory. Representation theory affords us a means to analyze the eigenvalues of Cayley graphs, at least for abelian groups.

**Definition 5.4.3 (Cayley graph).** Let $G$ be a finite group. By a *symmetric subset* of $G$, we mean a subset $S \subseteq G$ such that:

- $1 \notin S$;
- $s \in S$ implies $s^{-1} \in S$.

If $S$ is a symmetric subset of $G$, then the *Cayley graph* of $G$ with respect to $S$ is the graph with vertex set $G$ and with an edge $\{g, h\}$ connecting $g$ and $h$ if $gh^{-1} \in S$, or equivalently $hg^{-1} \in S$.

*Remark 5.4.4.* In this definition $S$ can be empty, in which case the Cayley graph has no edges. One can verify that the Cayley graph is connected (any two vertices can be connected by a path) if and only if $S$ generates $G$.

*Example 5.4.5.* Let $G = \mathbb{Z}/4\mathbb{Z}$ and $S = \{\pm[1]\}$. Then the Cayley graph of $G$ with respect to $S$ is drawn in Fig. 5.2. The adjacency matrix of this Cayley graph is given by

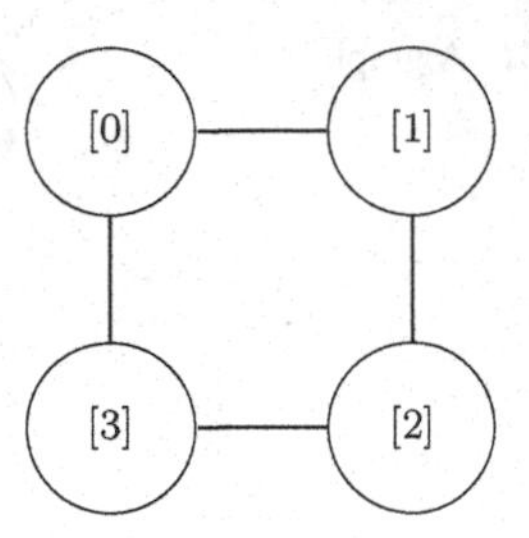


**Fig. 5.2** The Cayley graph of $\mathbb{Z}/4\mathbb{Z}$ with respect to $\{\pm[1]\}$

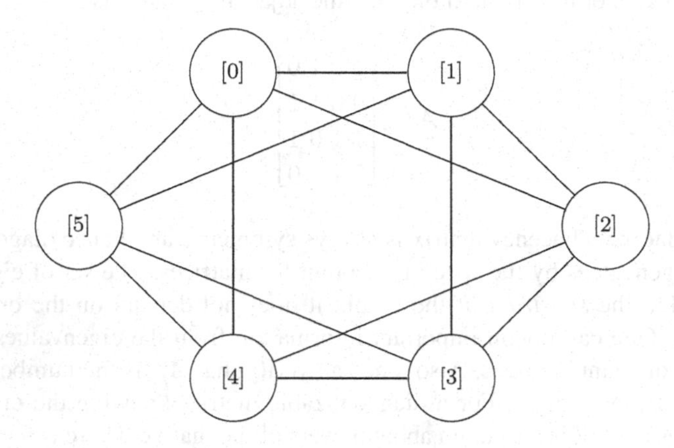


**Fig. 5.3** The Cayley graph of $\mathbb{Z}/6\mathbb{Z}$ with respect to $\{\pm[1], \pm[2]\}$

$$\begin{bmatrix} 0 & 1 & 0 & 1 \\ 1 & 0 & 1 & 0 \\ 0 & 1 & 0 & 1 \\ 1 & 0 & 1 & 0 \end{bmatrix}.$$

*Example 5.4.6.* In this example we take $G=\mathbb{Z}/6\mathbb{Z}$ and $S=\{\pm[1], \pm[2]\}$. The resulting Cayley graph can be found in Fig. 5.3. The adjacency matrix of this graph is

$$\begin{bmatrix} 0 & 1 & 1 & 0 & 1 & 1 \\ 1 & 0 & 1 & 1 & 0 & 1 \\ 1 & 1 & 0 & 1 & 1 & 0 \\ 0 & 1 & 1 & 0 & 1 & 1 \\ 1 & 0 & 1 & 1 & 0 & 1 \\ 1 & 1 & 0 & 1 & 1 & 0 \end{bmatrix}.$$

The graphs we have been considering are Cayley graphs of cyclic groups. Such graphs have a special name.

**Definition 5.4.7 (Circulant graph).** A Cayley graph of $\mathbb{Z}/n\mathbb{Z}$ is called a *circulant graph* (on $n$ vertices).

The adjacency matrix of a circulant graph is an example of a special type of matrix known as a circulant matrix.

**Definition 5.4.8 (Circulant matrix).** An $n \times n$ *circulant matrix* is a matrix of the form

$$A = \begin{bmatrix} a_0 & a_1 & \cdots & a_{n-2} & a_{n-1} \\ a_{n-1} & a_0 & a_1 & \cdots & a_{n-2} \\ \vdots & a_{n-1} & a_0 & \ddots & \vdots \\ a_2 & \vdots & \ddots & \ddots & a_1 \\ a_1 & a_2 & \cdots & a_{n-1} & a_0 \end{bmatrix}. \tag{5.5}$$

Equivalently, a matrix $A$ is circulant if there exists a function $f\colon \mathbb{Z}/n\mathbb{Z} \longrightarrow \mathbb{C}$ such that $A_{ij} = f([j] - [i])$. In (5.5), one has $a_i = f([i])$ for $0 \leq i \leq n-1$. If $S$ is a symmetric subset of $\mathbb{Z}/n\mathbb{Z}$, then the circulant matrix corresponding to the indicator functions $\delta_S$ of $S$ is the adjacency matrix of the Cayley graph of $\mathbb{Z}/n\mathbb{Z}$ with respect to $S$.

Our goal is to describe the eigenvalues of the Cayley graph of an abelian group. First we need a lemma about the group algebra $L(G)$.

**Lemma 5.4.9.** *Let $G$ be an abelian group and $a \in L(G)$. Define the convolution operator $A\colon L(G) \longrightarrow L(G)$ by $A(b) = a * b$. Then $A$ is linear and $\chi$ is an eigenvector of $A$ with eigenvalue $\widehat{a}(\chi)$ for all $\chi \in \widehat{G}$. Consequently, $A$ is a diagonalizable operator.*

*Proof.* Using the distributivity of convolution over addition, it is easy to verify that $A$ is linear. Let $n = |G|$ and suppose that $\chi \in \widehat{G}$. Observe that

$$\widehat{a * \chi} = \widehat{a} \cdot \widehat{\chi} = \widehat{a} \cdot n\delta_\chi$$

where the last equality uses Example 5.3.5. Clearly, one has that

$$(\widehat{a} \cdot n\delta_\chi)(\theta) = \begin{cases} \widehat{a}(\theta)n & \chi = \theta \\ 0 & \text{else} \end{cases}$$

for $\theta \in \widehat{G}$ and so $\widehat{a} \cdot n\delta_\chi = \widehat{a}(\chi)n\delta_\chi$. Applying the inverse of the Fourier transform to $\widehat{a * \chi} = \widehat{a}(\chi)n\delta_\chi$ and using that $\widehat{\chi} = n\delta_\chi$, we obtain $a * \chi = \widehat{a}(\chi)\chi$. In other words, $A\chi = \widehat{a}(\chi)\chi$ and so $\chi$ is an eigenvector of $A$ with eigenvalue $\widehat{a}(\chi)$.

Since the elements of $\widehat{G}$ form an orthonormal basis of eigenvectors for $A$, it follows that $A$ is diagonalizable. □

Lemma 5.4.9 is the key ingredient to computing the eigenvalues of the adjacency matrix of a Cayley graph of an abelian group. It only remains to realize the adjacency matrix as the matrix of a convolution operator.

**Theorem 5.4.10.** *Let $G = \{g_1, \ldots, g_n\}$ be an abelian group and $S \subseteq G$ a symmetric set. Let $\chi_1, \ldots, \chi_n$ be the irreducible characters of $G$ and let $A$ be the adjacency matrix of the Cayley graph of $G$ with respect to $S$ (using this ordering for the elements of $G$). Then:*

1. *The eigenvalues of the adjacency matrix $A$ are the real numbers*

$$\lambda_i = \sum_{s \in S} \chi_i(s)$$

*where $1 \leq i \leq n$;*

2. *The corresponding orthonormal basis of eigenvectors is given by the vectors $\{v_1, \ldots, v_n\}$ where*

$$v_i = \frac{1}{\sqrt{|G|}}(\chi_i(g_1), \ldots, \chi_i(g_n))^T.$$

*Proof.* Let $G = \{g_1, \ldots, g_n\}$ and let $\delta_S = \sum_{s \in S} \delta_s$ be the characteristic (or indicator) function of $S$; so

$$\delta_S(x) = \begin{cases} 1 & x \in S \\ 0 & \text{else.} \end{cases}$$

Let $F \colon L(G) \longrightarrow L(G)$ be the convolution operator

$$F(b) = \delta_S * b.$$

Lemma 5.4.9 implies that the irreducible characters $\chi_i$ are eigenvectors of $F$ and that the corresponding eigenvalue is

$$\widehat{\delta_S}(\chi_i) = n\langle \delta_S, \chi_i \rangle = \sum_{x \in S} \delta_S(x)\overline{\chi_i(x)} = \sum_{s \in S} \overline{\chi_i(s)} = \sum_{s \in S} \chi_i(s) = \lambda_i$$

where the penultimate equality is obtained by putting $s = x^{-1}$ and using that degree one representations are unitary, whence $\chi_i(x^{-1}) = \overline{\chi_i(x)}$, and that $S$ is symmetric.

It follows that if $B$ is the basis $\{\delta_{g_1}, \ldots, \delta_{g_n}\}$ for $L(G)$, then the matrix $[F]_B$ of $F$ with respect to this basis has eigenvalues $\lambda_1, \ldots, \lambda_n$ and eigenvectors $v_1, \ldots, v_n$. The orthonormality of the $v_i$ follows from the orthonormality of the $\chi_i$; the scaling by $1/\sqrt{|G|}$ comes from the fact that the $\delta_{g_i}$ are orthonormal with respect to the inner product $(f_1, f_2) = |G|\langle f_1, f_2 \rangle$. Therefore, it remains to prove that $A = [F]_B$.

To this end we compute

$$F(\delta_{g_j}) = \delta_S * \delta_{g_j} = \sum_{s\in S} \delta_s * \delta_{g_j} = \sum_{s\in S} \delta_{sg_j}$$

by Proposition 5.2.2. Recalling that $([F]_B)_{ij}$ is the coefficient of $\delta_{g_i}$ in $F(\delta_{g_j})$, we conclude that

$$\begin{aligned}([F]_B)_{ij} &= \begin{cases} 1 & g_i = sg_j \text{ for some } s \in S \\ 0 & \text{else} \end{cases} \\ &= \begin{cases} 1 & g_i g_j^{-1} \in S \\ 0 & \text{else} \end{cases} \\ &= A_{ij}\end{aligned}$$

as required.

Finally, to verify that $\lambda_i$ is real, we just observe that if $s \in S$, then either $s = s^{-1}$, and so $\chi_i(s) = \chi_i(s^{-1}) = \overline{\chi_i(s)}$ is real, or $s \neq s^{-1} \in S$ and $\chi(s) + \chi(s^{-1}) = \chi(s) + \overline{\chi(s)}$ is real. □

Specializing to the case of circulant matrices, we obtain:

**Corollary 5.4.11.** *Let $A$ be a circulant matrix of degree $n$, which is the adjacency matrix of the Cayley graph of $\mathbb{Z}/n\mathbb{Z}$ with respect to the symmetric set $S$. Then the eigenvalues of $A$ are*

$$\lambda_k = \sum_{[m]\in S} e^{2\pi ikm/n}$$

*where $k = 0, \ldots, n-1$ and a corresponding basis of orthonormal eigenvectors is given by $v_0, \ldots, v_{n-1}$ where*

$$v_k = \frac{1}{\sqrt{n}}(1, e^{2\pi ik2/n}, \ldots, e^{2\pi ik(n-1)/n})^T.$$

*Example 5.4.12.* Let $A$ be the adjacency matrix of the circulant graph in Example 5.4.6. Then the eigenvalues of $A$ are $\lambda_1, \ldots, \lambda_6$ where

$$\lambda_k = e^{\pi ik/3} + e^{-\pi ik/3} + e^{2\pi ik/3} + e^{-2\pi ik/3} = 2\cos \pi k/3 + 2\cos 2\pi k/3$$

for $k = 1, \ldots, 6$.

*Remark 5.4.13.* This approach can be generalized to non-abelian groups provided the symmetric set $S$ is closed under conjugation. For more on the relationship between graph theory and representation theory, as well as the related subject of random walks on graphs, see [3, 6, 7].

## 5.5 Fourier Analysis on Non-abelian Groups

For a non-abelian group $G$, we have $L(G) \neq Z(L(G))$ and so $L(G)$ is a non-commutative ring. Therefore, we cannot find a Fourier transform that turns convolution into pointwise multiplication (as pointwise multiplication is commutative). Instead, we try to replace pointwise multiplication by matrix multiplication. To achieve this, let us first recast the abelian case in a different form.

Suppose that $G$ is a finite abelian group with irreducible characters $\chi_1, \ldots, \chi_n$. Then to each function $f\colon G \longrightarrow \mathbb{C}$, we can associate its vector of Fourier coefficients. That is, we define $T\colon L(G) \longrightarrow \mathbb{C}^n$ by

$$Tf = (n\langle f, \chi_1\rangle, n\langle f, \chi_2\rangle, \ldots, n\langle f, \chi_n\rangle) = (\widehat{f}(\chi_1), \widehat{f}(\chi_2), \ldots, \widehat{f}(\chi_n)).$$

The map $T$ is injective by the Fourier inversion theorem because we can recover $\widehat{f}$, and hence $f$, from $Tf$. It is also linear (this is essentially a reformulation of Proposition 5.3.7) and hence a vector space isomorphism since $\dim L(G) = n$. Now $\mathbb{C}^n = \mathbb{C} \times \cdots \times \mathbb{C}$ has the structure of a direct product of rings where multiplication is taken coordinate-wise:

$$(a_1, \ldots, a_n)(b_1, \ldots, b_n) = (a_1 b_1, \ldots, a_n b_n).$$

The map $T$ is in fact a ring isomorphism since

$$\begin{aligned} T(a * b) &= (\widehat{a * b}(\chi_1), \ldots, \widehat{a * b}(\chi_n)) = (\widehat{a}(\chi_1)\widehat{b}(\chi_1), \ldots, \widehat{a}(\chi_n)\widehat{b}(\chi_n)) \\ &= (\widehat{a}(\chi_1), \ldots, \widehat{a}(\chi_n))(\widehat{b}(\chi_1), \ldots, \widehat{b}(\chi_n)) = Ta \cdot Tb \end{aligned}$$

Theorem 5.3.8 can thus be reinterpreted in the following way.

**Theorem 5.5.1.** *Let $G$ be a finite abelian group of order $n$. Then $L(G) \cong \mathbb{C}^n$.*

One might guess that this reflects the fact that all irreducible representations of an abelian group have degree one and that, for non-abelian groups, we must replace $\mathbb{C}$ by matrix rings over $\mathbb{C}$. This is indeed the case. So without further ado, let $G$ be a finite group of order $n$ with complete set $\varphi^{(1)}, \ldots, \varphi^{(s)}$ of unitary representatives of the equivalence classes of irreducible representations of $G$. As usual, we put $d_k = \deg \varphi^{(k)}$. The matrix coefficients are the functions $\varphi_{ij}^{(k)}\colon G \longrightarrow \mathbb{C}$ given by $\varphi_g^{(k)} = (\varphi_{ij}^{(k)}(g))$. Theorem 4.4.6 tells us that the functions $\sqrt{d_k}\varphi_{ij}^{(k)}$ form an orthonormal basis for $L(G)$.

**Definition 5.5.2 (Fourier transform).** Define

$$T\colon L(G) \longrightarrow M_{d_1}(\mathbb{C}) \times \cdots \times M_{d_s}(\mathbb{C})$$

by $Tf = (\widehat{f}(\varphi^{(1)}), \ldots, \widehat{f}(\varphi^{(s)}))$ where

$$\widehat{f}(\varphi^{(k)})_{ij} = n\langle f, \varphi_{ij}^{(k)}\rangle = \sum_{g \in G} f(g)\overline{\varphi_{ij}^{(k)}(g)}. \tag{5.6}$$

We call $Tf$ the *Fourier transform* of $f$.

Notice that (5.6) can be written more succinctly in the form

$$\widehat{f}(\varphi^{(k)}) = \sum_{g\in G} f(g)\overline{\varphi_g^{(k)}}$$

which is the form that we shall most frequently use.

*Remark 5.5.3.* Many authors define

$$\widehat{f}(\varphi^{(k)}) = \sum_{g\in G} f(g)\varphi_g^{(k)}. \tag{5.7}$$

Our convention was chosen for consistency with the case of abelian groups, where it is well-entrenched. Having said that, there are a number of advantages to choosing (5.7) as the definition of the Fourier transform, and perhaps future generations will switch the convention for abelian groups.

Let us begin our study of non-abelian Fourier analysis with the Fourier inversion theorem.

**Theorem 5.5.4 (Fourier inversion).** *Let $f\colon G \longrightarrow \mathbb{C}$ be a complex-valued function on G. Then*

$$f = \frac{1}{n}\sum_{i,j,k} d_k \widehat{f}(\varphi^{(k)})_{ij}\varphi_{ij}^{(k)}$$

*where $n = |G|$.*

*Proof.* We compute using the orthonormality of the $\sqrt{d_k}\varphi_{ij}^{(k)}$

$$\begin{aligned} f &= \sum_{i,j,k}\left\langle f, \sqrt{d_k}\varphi_{ij}^{(k)}\right\rangle \sqrt{d_k}\varphi_{ij}^{(k)} = \frac{1}{n}\sum_{i,j,k} d_k n \left\langle f, \varphi_{ij}^{(k)}\right\rangle \varphi_{ij}^{(k)} \\ &= \frac{1}{n}\sum_{i,j,k} d_k \widehat{f}(\varphi^{(k)})_{ij}\varphi_{ij}^{(k)} \end{aligned}$$

as required. □

Next we show that $T$ is a vector space isomorphism.

**Proposition 5.5.5.** *The map $T\colon L(G) \longrightarrow M_{d_1}(\mathbb{C}) \times \cdots \times M_{d_s}(\mathbb{C})$ is a vector space isomorphism.*

*Proof.* To show that $T$ is linear it suffices to prove

$$\left(\widehat{c_1 f_1 + c_2 f_2}\right)(\varphi^{(k)}) = c_1\widehat{f_1}(\varphi^{(k)}) + c_2\widehat{f_2}(\varphi^{(k)})$$

for $1 \leq k \leq s$. Indeed,

$$\begin{aligned}
\left(\widehat{c_1 f_1 + c_2 f_2}\right)(\varphi^{(k)}) &= \sum_{g\in G} (c_1 f_1 + c_2 f_2)(g)\overline{\varphi_g^{(k)}} \\
&= c_1 \sum_{g\in G} f_1(g)\overline{\varphi_g^{(k)}} + c_2 \sum_{g\in G} f_2(g)\overline{\varphi_g^{(k)}} \\
&= c_1 \widehat{f_1}(\varphi^{(k)}) + c_2 \widehat{f_2}(\varphi^{(k)})
\end{aligned}$$

as was to be proved.

The Fourier inversion theorem implies that $T$ is injective. Since

$$\dim L(G) = |G| = d_1^2 + \cdots + d_s^2 = \dim M_{d_1}(\mathbb{C}) \times \cdots \times M_{d_s}(\mathbb{C})$$

it follows that $T$ is an isomorphism. $\square$

All the preparation has now been completed to show that the Fourier transform is a ring isomorphism. This leads us to a special case of a more general theorem of Wedderburn that is often taken as the starting point for studying the representation theory of finite groups.

**Theorem 5.5.6 (Wedderburn).** *The Fourier transform*

$$T\colon L(G) \longrightarrow M_{d_1}(\mathbb{C}) \times \cdots \times M_{d_s}(\mathbb{C})$$

*is an isomorphism of rings.*

*Proof.* Proposition 5.5.5 asserts that $T$ is an isomorphism of vector spaces. Therefore, to show that it is a ring isomorphism it suffices to verify that $T(a*b) = Ta \cdot Tb$. In turn, by the definition of multiplication in a direct product, to do this it suffices to establish $\widehat{a*b}(\varphi^{(k)}) = \widehat{a}(\varphi^{(k)}) \cdot \widehat{b}(\varphi^{(k)})$ for $1 \leq k \leq s$. The computation is analogous to the abelian case:

$$\begin{aligned}
\widehat{a*b}(\varphi^{(k)}) &= \sum_{x\in G} (a*b)(x)\overline{\varphi_x^{(k)}} \\
&= \sum_{x\in G} \overline{\varphi_x^{(k)}} \sum_{y\in G} a(xy^{-1})b(y) \\
&= \sum_{y\in G} b(y) \sum_{x\in G} a(xy^{-1})\overline{\varphi_x^{(k)}}.
\end{aligned}$$

Setting $z = xy^{-1}$ (and so $x = zy$) yields

$$\begin{aligned}\widehat{a * b}(\varphi^{(k)}) &= \sum_{y \in G} b(y) \sum_{z \in G} a(z) \overline{\varphi_{zy}^{(k)}} \\ &= \sum_{y \in G} b(y) \sum_{z \in G} a(z) \overline{\varphi_z^{(k)}} \cdot \overline{\varphi_y^{(k)}} \\ &= \sum_{z \in G} a(z) \overline{\varphi_z^{(k)}} \sum_{y \in G} b(y) \overline{\varphi_y^{(k)}} \\ &= \widehat{a}(\varphi^{(k)}) \cdot \widehat{b}(\varphi^{(k)})\end{aligned}$$

This concludes the proof that $T$ is a ring isomorphism. □

For non-abelian groups, it is still true that computing $Ta \cdot Tb$ and inverting $T$ can sometimes be faster than computing $a * b$ directly.

*Remark 5.5.7.* Note that

$$\widehat{\delta_g}(\varphi^{(k)}) = \sum_{x \in G} \delta_g(x) \overline{\varphi_x^{(k)}} = \overline{\varphi_g^{(k)}}.$$

Since the conjugate of an irreducible representation is easily verified to be irreducible, it follows that $T\delta_g$ is a vector whose entries consist of the images of $g$ under all the irreducible representations of $G$, in some order.

The next example gives some indication how the representation theory of $S_n$ can be used to analyze voting.

*Example 5.5.8 (Diaconis).* Suppose that in an election each voter has to rank $n$ candidates on a ballot. Let us call the candidates $\{1, \ldots, n\}$. Then to each ballot we can correspond a permutation $\sigma \in S_n$. For example, if $n = 3$ and the ballot ranks the candidates in the order 312, then the corresponding permutation is

$$\sigma = \begin{pmatrix} 1 & 2 & 3 \\ 3 & 1 & 2 \end{pmatrix}.$$

An election then corresponds to a function $f \colon S_n \longrightarrow \mathbb{N}$ where $f(\sigma)$ is the number of people whose ballot corresponds to the permutation $\sigma$. Using the fast Fourier transform for the symmetric group, Diaconis was able to analyze various elections. As with signal processing, one may discard Fourier coefficients of small magnitude to compress data. Also for $S_n$, the Fourier coefficients $n!\langle f, \varphi_{ij}^{(k)} \rangle$ have nice interpretations. For instance, an appropriate coefficient measures how many people ranked candidate $m$ first among all candidates. See [7, 8] for details.

## Exercises

**Exercise 5.1.** Prove that if $a, b\colon G \longrightarrow \mathbb{C}$, then

$$a * b(x) = \sum_{y \in G} a(y)b(y^{-1}x).$$

**Exercise 5.2.** Verify that $L(G)$ satisfies the distributive laws.

**Exercise 5.3.** Let $f\colon \mathbb{Z}/3\mathbb{Z} \longrightarrow \mathbb{C}$ be given by $f([k]) = \sin(2\pi k/3)$. Compute the Fourier transform $\widehat{f}$ of $f$ (identifying $\widehat{\mathbb{Z}/3\mathbb{Z}}$ with $\mathbb{Z}/3\mathbb{Z}$ in the usual way).

**Exercise 5.4.** Draw the Cayley graph of $\mathbb{Z}/6\mathbb{Z}$ with respect to the set $S = \{\pm[2], \pm[3]\}$ and compute the eigenvalues of the adjacency matrix.

**Exercise 5.5.** Draw the Cayley graph of $(\mathbb{Z}/2\mathbb{Z})^3$ with respect to the set $S = \{([1],[0],[0]), ([0],[1],[0]), ([0],[0],[1])\}$ and compute the eigenvalues of the adjacency matrix.

**Exercise 5.6.** Let $G$ be a finite abelian group of order $n$ and $a, b \in L(G)$. Prove the Plancherel formula

$$\langle a, b\rangle = \frac{1}{n}\langle \widehat{a}, \widehat{b}\rangle.$$

**Exercise 5.7.** Let $G$ be a finite group of order $n$ and let $a, b \in L(G)$. Suppose that $\varphi^{(1)}, \ldots, \varphi^{(s)}$ form a complete set of representatives of the irreducible unitary representations of $G$. As usual, let $d_i$ be the degree of $\varphi^{(i)}$. Prove the Plancherel formula

$$\langle a, b\rangle = \frac{1}{n^2}\sum_{i=1}^{s} d_i \operatorname{Tr}\left[\widehat{a}(\varphi^{(i)})\widehat{b}(\varphi^{(i)})^*\right].$$

**Exercise 5.8.** Prove Lemma 5.4.9 directly from the definition of convolution.

**Exercise 5.9.** Let $G$ be a finite abelian group and let $a \in L(G)$. Let $A$ be the convolution operator $A\colon L(G) \longrightarrow L(G)$ defined by $A(b) = a * b$. Show that the following are equivalent:

1. $A$ is self-adjoint;
2. The eigenvalues of $A$ are real;
3. $\widehat{a}(\chi)$ is real for all $\chi \in \widehat{G}$;
4. $a(g) = \overline{a(g^{-1})}$ for all $g \in G$.

(Hint: you might find Exercise 2.6 useful.)

**Exercise 5.10.** Prove that $Z(M_n(\mathbb{C})) = \{\lambda I \mid \lambda \in \mathbb{C}\}$.

**Exercise 5.11.** Let $G$ be a finite group of order $n$ and let $\varphi^{(1)}, \ldots, \varphi^{(s)}$ be a complete set of unitary representatives of the equivalence classes of irreducible representations of $G$. Let $\chi_i$ be the character of $\varphi^{(i)}$ and let $e_i = \frac{d_i}{n}\chi_i$ where $d_i$ is the degree of $\varphi^{(i)}$.

1. Show that if $f \in Z(L(G))$, then

$$\widehat{f}(\varphi^{(k)}) = \frac{n}{d_k}\langle f, \chi_k\rangle I.$$

2. Deduce that

$$\widehat{e_i}(\varphi^{(k)}) = \begin{cases} I & i = k \\ 0 & \text{else.} \end{cases}$$

3. Deduce that

$$e_i * e_j = \begin{cases} e_i & i = j \\ 0 & \text{else.} \end{cases}$$

4. Deduce that $e_1 + \cdots + e_s$ is the identity $\delta_1$ of $L(G)$.

**Exercise 5.12.** Let $G = \{g_1, \ldots, g_n\}$ be a finite group of order $n$ and let $\varphi^{(1)}, \ldots, \varphi^{(s)}$ be a complete set of unitary representatives of the equivalence classes of irreducible representations of $G$. Let $\chi_i$ be the character of $\varphi^{(i)}$ and $d_i$ be the degree of $\varphi^{(i)}$. Suppose $a \in Z(L(G))$ and define a linear operator $A\colon L(G) \longrightarrow L(G)$ by $A(b) = a * b$.

1. Fix $1 \leq k \leq s$. Show that $\varphi_{ij}^{(k)}$ is an eigenvector of $A$ with eigenvalue $\frac{n}{d_k}\langle a, \chi_k\rangle$. (Hint: show that

$$\widehat{\varphi_{ij}^{(m)}}(\varphi^{(k)}) = \begin{cases} \dfrac{n}{d_k}E_{ij} & m = k \\ 0 & \text{else.} \end{cases}$$

   Now compute $\widehat{a * \varphi_{ij}^{(k)}}$ using Exercise 5.11(1) and apply the Fourier inversion theorem.)
2. Conclude that $A$ is a diagonalizable operator.
3. Let $S \subseteq G$ be a symmetric set and assume further that $gSg^{-1} = S$ for all $g \in G$. Show that the eigenvalues of the adjacency matrix $A$ of the Cayley graph of $G$ with respect to $S$ are $\lambda_1, \ldots, \lambda_s$ where

$$\lambda_k = \frac{1}{d_k}\sum_{s \in S}\chi_k(s)$$

and that $\lambda_k$ has multiplicity $d_k^2$. Show that the vectors

$$v_{ij}^{(k)} = \frac{\sqrt{d_k}}{\sqrt{|G|}}(\varphi_{ij}^{(k)}(g_1), \ldots, \varphi_{ij}^{(k)}(g_n))^T$$

with $1 \leq i, j \leq d_k$ form an orthonormal basis for the eigenspace $V_{\lambda_k}$.

4. Compute the eigenvalues of the Cayley graph of $S_3$ with respect to $S = \{(1\ 2), (1\ 3), (2\ 3)\}$.

**Exercise 5.13.** Let $G$ be a finite abelian group.

1. Define $\eta\colon G \longrightarrow \widehat{\widehat{G}}$ by $\eta(g)(\chi) = \chi(g)$. Prove that $\eta$ is an isomorphism.
2. Prove that $G \cong \widehat{G}$.

**Exercise 5.14.** Let $f\colon \mathbb{Z}/n\mathbb{Z} \longrightarrow \mathbb{C}$ be a function and let $A$ be the corresponding circulant matrix defined in Definition 5.4.8. Let $\chi_0, \ldots, \chi_{n-1}$ be as in Example 5.3.3.

1. Show that the eigenvalues of $A$ are given by $\lambda_i = \widehat{f}(\chi_i)$ for $i = 0, \ldots, n-1$.
2. Show that $\{v_0, \ldots, v_{n-1}\}$ given by

$$v_k = \frac{1}{\sqrt{n}}(1, e^{2\pi ik2/n}, \ldots, e^{2\pi ik(n-1)/n})^T$$

is an orthonormal basis of eigenvectors for $A$ with $v_i$ having eigenvalue $\lambda_i$.

3. Deduce that $A$ is diagonalizable.

(Hint: show that $A$ is the matrix of the linear mapping $a \mapsto f * a$ with respect to the basis $\{\delta_{[0]}, \ldots, \delta_{[n-1]}\}$ of $L(\mathbb{Z}/n\mathbb{Z})$.)

**Exercise 5.15.** Prove that the set of $n \times n$ circulant matrices is a subring of $M_n(\mathbb{C})$ isomorphic to $L(\mathbb{Z}/n\mathbb{Z})$.

**Exercise 5.16.** The following exercise is for readers familiar with probability and statistics. Let $G$ be a finite group and suppose that $X, Y$ are random variables taking values in $G$ with distributions $\mu, \nu$ respectively, that is,

$$\text{Prob}[X = g] = \mu(g) \quad \text{and} \quad \text{Prob}[Y = g] = \nu(g)$$

for $g \in G$. Show that if $X$ and $Y$ are independent, then the random variable $XY$ has distribution the convolution $\mu * \nu$. Thus the Fourier transform is useful for studying products of group-valued random variables [7].

# Chapter 6
# Burnside's Theorem

In this chapter, we look at one of the first major applications of representation theory: Burnside's $pq$-theorem. This theorem states that no non-abelian group of order $p^a q^b$ is simple. Recall that a group is *simple* if it contains no non-trivial proper normal subgroups. It took nearly seventy years (cf. [2,14]) to find a proof that avoids representation theory! To prove Burnside's theorem we shall need to take a brief excursion into number theory. On the way, we shall prove a result of Frobenius, sometimes known as the dimension theorem, which says that the degree of each irreducible representation of a group $G$ divides the order of $G$. This fact turns out to be quite useful for determining the character table of a group.

## 6.1 A Little Number Theory

A complex number is called an *algebraic number* if it is the root of a polynomial with integer coefficients. Numbers that are not algebraic are called *transcendental*. For instance $1/2$ is algebraic, being a root of the polynomial $2z - 1$, and so is $\sqrt{2}$, as it is a root of $z^2 - 2$. A standard course in rings and fields shows that the set $\overline{\mathbb{Q}}$ of algebraic numbers is a field (cf. [11]). A fairly straightforward counting argument shows that $\overline{\mathbb{Q}}$ is countable (there are only countably many polynomials over $\mathbb{Z}$ and each one has only finitely many roots). Hopefully, the reader has already encountered the fact that $\mathbb{C}$ is uncountable. Thus most numbers are not algebraic. However, it is extremely difficult to prove that a given number is transcendental. For example $e$ and $\pi$ are transcendental, but this is highly non-trivial to prove. Although number theory is concerned with numbers in general, it is primarily focused on integers and so for our purposes we are interested in a special type of algebraic number called an algebraic integer.

B. Steinberg, *Representation Theory of Finite Groups: An Introductory Approach*, Universitext, DOI 10.1007/978-1-4614-0776-8_6,
© Springer Science+Business Media, LLC 2012

**Definition 6.1.1 (Algebraic integer).** A complex number $\alpha$ is said to be an *algebraic integer* if it is a root of a monic polynomial with integer coefficients. That is to say, $\alpha$ is an algebraic integer if there is a polynomial

$$p(z) = z^n + a_{n-1}z^{n-1} + \cdots + a_0$$

with $a_0, \ldots, a_{n-1} \in \mathbb{Z}$ and $p(\alpha) = 0$.

The fact that the leading coefficient of $p$ is 1 is crucial to the definition. Trivially, every integer $m \in \mathbb{Z}$ is an algebraic integer, being the root of the polynomial $z - m$. Notice that if $\alpha$ is an algebraic integer, then so is $-\alpha$ because if $p(z)$ is a monic polynomial with integer coefficients such that $p(\alpha) = 0$, then either $p(-z)$ or $-p(-z)$ is a monic polynomial and $-\alpha$ is a root of both these polynomials. In fact, we shall see later that the algebraic integers form a subring of $\mathbb{C}$.

*Example 6.1.2 ($n^{th}$-roots).* Let $m$ be an integer. Then $z^n - m$ is a monic polynomial with integer coefficients, so any $n$th-root of $m$ is an algebraic integer. Thus, for example, $\sqrt{2}$ is an algebraic integer, as is $e^{2\pi i/n}$. In fact, any $n$th-root of unity is an algebraic integer.

*Example 6.1.3 (Eigenvalues of integer matrices).* Let $A = (a_{ij})$ with the $a_{ij} \in \mathbb{Z}$ be an $n \times n$ integer matrix. Then the characteristic polynomial $p_A(z) = \det(zI - A)$ is a monic polynomial with integer coefficients. Thus each eigenvalue of $A$ is an algebraic integer.

A rational number $m/n$ is a root of the non-monic integral polynomial $nz - m$. One would guess that a rational number ought not be an algebraic integer unless it is in fact an integer. This is indeed the case, as follows from the "rational roots test" from high school.

**Proposition 6.1.4.** *A rational number is an algebraic integer if and only if it is an integer.*

*Proof.* Let $r = m/n$ with $m, n \in \mathbb{Z}$, $n > 0$ and $\gcd(m, n) = 1$. Suppose that $r$ is a root of the polynomial with integer coefficients $z^k + a_{k-1}z^{k-1} + \cdots + a_0$. Then

$$0 = \left(\frac{m}{n}\right)^k + a_{k-1}\left(\frac{m}{n}\right)^{k-1} + \cdots + a_0$$

and so clearing denominators (by multiplying by $n^k$) yields

$$0 = m^k + a_{k-1}m^{k-1}n + \cdots + a_1mn^{k-1} + a_0n^k.$$

In other words,

$$m^k = -n(a_{k-1}m^{k-1} + \cdots + a_1mn^{k-1} + a_0n^{k-1})$$

and so $n \mid m^k$. As $\gcd(m, n) = 1$, we conclude $n = \pm 1$. Thus $r = m \in \mathbb{Z}$. □

A general strategy we shall employ to show that an integer $d$ divides an integer $n$ is to show that $n/d$ is an algebraic integer. Proposition 6.1.4 then implies $d \mid n$. First we need to learn more about algebraic integers. Namely, we want to show that they form a subring $\mathbb{A}$ of $\mathbb{C}$. To do this we need the following characterization of algebraic integers.

**Lemma 6.1.5.** *An element $y \in \mathbb{C}$ is an algebraic integer if and only if there exist $y_1, \ldots, y_t \in \mathbb{C}$, not all zero, such that*

$$yy_i = \sum_{j=1}^{t} a_{ij}y_j$$

*with the $a_{ij} \in \mathbb{Z}$ for all $1 \leq i \leq t$ (i.e., $yy_i$ is an integral linear combination of the $y_i$ for all $i$).*

*Proof.* Suppose first that $y$ is an algebraic integer. Let $y$ be a root of

$$p(z) = z^n + a_{n-1}z^{n-1} + \cdots + a_0$$

and take $y_i = y^{i-1}$ for $1 \leq i \leq n$. Then, for $1 \leq i \leq n-2$, we have $yy_i = yy^{i-1} = y^i = y_{i+1}$ and $yy_{n-1} = y^n = -a_0 - \cdots - a_{n-1}y^{n-1}$.

Conversely, if $y_1, \ldots, y_t$ are as in the statement of the lemma, let $A = (a_{ij})$ and

$$Y = \begin{bmatrix} y_1 \\ y_2 \\ \vdots \\ y_t \end{bmatrix} \in \mathbb{C}^t.$$

Then

$$[AY]_i = \sum_{j=1}^{t} a_{ij}y_j = yy_i = y[Y]_i$$

and so $AY = yY$. Since $Y \neq 0$ by assumption, it follows that $y$ is an eigenvalue of the $t \times t$ integer matrix $A$ and hence is an algebraic integer by Example 6.1.3. □

**Corollary 6.1.6.** *The set $\mathbb{A}$ of algebraic integers is a subring of $\mathbb{C}$. In particular, the sum and product of algebraic integers is algebraic.*

*Proof.* We already observed that $\mathbb{A}$ is closed under taking negatives. Let $y, y' \in \mathbb{A}$. Choose $y_1, y_2, \ldots, y_t \in \mathbb{C}$ not all 0 and $y'_1, \ldots, y'_s \in \mathbb{C}$ not all 0 such that

$$yy_i = \sum_{j=1}^{t} a_{ij}y_j, \; y'y'_k = \sum_{j=1}^{s} b_{kj}y'_j$$

as guaranteed by Lemma 6.1.5. Then

$$(y+y')y_iy'_k = yy_iy'_k + y'y'_ky_i = \sum_{j=1}^{t} a_{ij}y_jy'_k + \sum_{j=1}^{s} b_{kj}y'_jy_i$$

is an integral linear combination of the $y_jy'_\ell$, establishing that $y+y' \in \mathbb{A}$ by Lemma 6.1.5. Similarly, $yy'y_iy'_k = yy_iy'y'_k$ is an integral linear combination of the $y_jy'_\ell$ and so $yy' \in \mathbb{A}$. □

We shall also need that the complex conjugate of an algebraic integer is an algebraic integer. Indeed, if $p(z) = z^n + a_{n-1}z^{n-1} + \cdots + a_0$ is a polynomial with integer coefficients and $\alpha$ is a root of $p(z)$, then

$$p(\overline{\alpha}) = \overline{\alpha}^n + a_{n-1}\overline{\alpha}^{n-1} + \cdots + a_0 = \overline{\alpha^n + a_{n-1}\alpha^{n-1} + \cdots + a_0} = \overline{p(\alpha)} = 0.$$

## 6.2 The Dimension Theorem

The relevance of algebraic integers to group representation theory becomes apparent by considering the following consequence to Corollary 6.1.6.

**Corollary 6.2.1.** *Let $\chi$ be a character of a finite group $G$. Then $\chi(g)$ is an algebraic integer for all $g \in G$.*

*Proof.* Let $\varphi\colon G \longrightarrow GL_m(\mathbb{C})$ be a representation with character $\chi$ and let $n$ be the order of $G$. Then $g^n = 1$ and so $\varphi_g^n = I$. Corollary 4.1.10 then implies that $\varphi_g$ is diagonalizable with eigenvalues $\lambda_1, \ldots, \lambda_m$, which are $n$th-roots of unity. In particular, the eigenvalues of $\varphi_g$ are algebraic integers. Since

$$\chi(g) = \mathrm{Tr}(\varphi_g) = \lambda_1 + \cdots + \lambda_m$$

and the algebraic integers form a ring, we conclude that $\chi(g)$ is an algebraic integer. □

*Remark 6.2.2.* Notice that the proof of Corollary 6.2.1 shows that $\chi_\varphi(g)$ is a sum of $m$ $n$th-roots of unity. We shall use this fact later.

Our next goal is to show that the degree of an irreducible representation divides the order of the group. To do this we need to conjure up some more algebraic integers.

**Theorem 6.2.3.** *Let $\varphi$ be an irreducible representation of a finite group $G$ of degree $d$. Let $g \in G$ and let $h$ be the size of the conjugacy class of $g$. Then $h\chi_\varphi(g)/d$ is an algebraic integer.*

*Proof.* Let $C_1, \ldots, C_s$ be the conjugacy classes of $G$. Set $h_i = |C_i|$ and let $\chi_i$ be the value of $\chi_\varphi$ on the class $C_i$. We want to show that $h_i\chi_i/d$ is an algebraic integer for each $i$. Consider the operator

$$T_i = \sum_{x\in C_i} \varphi_x.$$

*Claim.* $T_i = \frac{h_i}{d}\chi_i \cdot I$.

*Proof (of claim).* We first show that $\varphi_g T_i \varphi_{g^{-1}} = T_i$ for all $g \in G$. Indeed,

$$\varphi_g T_i \varphi_{g^{-1}} = \sum_{x\in C_i} \varphi_g\varphi_x\varphi_{g^{-1}} = \sum_{x\in C_i} \varphi_{gxg^{-1}} = \sum_{y\in C_i} \varphi_y = T_i$$

since $C_i$ is closed under conjugation and conjugation by $g$ is a permutation. By Schur's lemma, $T_i = \lambda I$ some $\lambda \in \mathbb{C}$. Then since $I$ is the identity operator on a $d$-dimensional vector space, it follows that

$$d\lambda = \operatorname{Tr}(\lambda I) = \operatorname{Tr}(T_i) = \sum_{x\in C_i} \operatorname{Tr}(\varphi_x) = \sum_{x\in C_i} \chi_\varphi(x) = \sum_{x\in C_i} \chi_i = |C_i|\chi_i = h_i\chi_i$$

and so $\lambda = h_i\chi_i/d$, establishing the claim. □

We need yet another claim, which says that the $T_i$ "behave" like algebraic integers in that they satisfy a formula like in Lemma 6.1.5.

*Claim.* $T_iT_j = \sum_{k=1}^{s} a_{ijk}T_k$ for some $a_{ijk} \in \mathbb{Z}$.

*Proof (of claim).* Routine calculation shows

$$T_iT_j = \sum_{x\in C_i} \varphi_x \cdot \sum_{y\in C_j} \varphi_y = \sum_{x\in C_i, y\in C_j} \varphi_{xy} = \sum_{g\in G} a_{ijg}\varphi_g$$

where $a_{ijg} \in \mathbb{Z}$ is the number of ways to write $g = xy$ with $x \in C_i$ and $y \in C_j$. We claim that $a_{ijg}$ depends only on the conjugacy class of $g$. Suppose that this is indeed the case and let $a_{ijk}$ be the value of $a_{ijg}$ with $g \in C_k$. Then

$$\sum_{g\in G} a_{ijg}\varphi_g = \sum_{k=1}^{s}\sum_{g\in C_k} a_{ijg}\varphi_g = \sum_{k=1}^{s} a_{ijk}\sum_{g\in C_k}\varphi_g = \sum_{k=1}^{s} a_{ijk}T_k$$

proving the claim.

So let us check that $a_{ijg}$ depends only on the conjugacy class of $g$. Let

$$X_g = \{(x,y) \in C_i \times C_j \mid xy = g\};$$

so $a_{ijg} = |X_g|$. Let $g'$ be conjugate to $g$. We show that $|X_g| = |X_{g'}|$. Suppose that $g' = kgk^{-1}$ and define a bijection $\psi\colon X_g \longrightarrow X'_g$ by

$$\psi(x, y) = (kxk^{-1}, kyk^{-1}).$$

Notice that $kxk^{-1} \in C_i$, $kyk^{-1} \in C_j$ and $kxk^{-1}kyk^{-1} = kxyk^{-1} = kgk^{-1} = g'$, and so $\psi(x, y) \in X'_g$. Evidently, $\psi$ has inverse $\tau\colon X_{g'} \longrightarrow X_g$ given by $\tau(x', y') = (k^{-1}x'k, k^{-1}y'k)$ so $\psi$ is a bijection and hence $|X_g| = |X_{g'}|$. □

We now complete the proof of the theorem. Substituting the formula for the $T_i$ from the first claim into the formula from the second claim yields

$$\left(\frac{h_i}{d}\chi_i\right) \cdot \left(\frac{h_j}{d}\chi_j\right) = \sum_{k=1}^{s} a_{ijk} \left(\frac{h_k}{d}\chi_k\right)$$

and so $h_i\chi_i/d$ is an algebraic integer by Lemma 6.1.5. □

**Theorem 6.2.4 (Dimension theorem).** *Let $\varphi$ be an irreducible representation of $G$ of degree $d$. Then $d$ divides $|G|$.*

*Proof.* The first orthogonality relations (Theorem 4.3.9) provide

$$1 = \langle \chi_\varphi, \chi_\varphi \rangle = \frac{1}{|G|} \sum_{g \in G} \chi_\varphi(g)\overline{\chi_\varphi(g)},$$

and so

$$\frac{|G|}{d} = \sum_{g \in G} \frac{\chi_\varphi(g)}{d} \overline{\chi_\varphi(g)}. \tag{6.1}$$

Let $C_1, \ldots, C_s$ be the conjugacy classes of $G$ and let $\chi_i$ be the value of $\chi_\varphi$ on $C_i$. Let $h_i = |C_i|$. Then from (6.1) we obtain

$$\frac{|G|}{d} = \sum_{i=1}^{s} \sum_{g \in C_i} \frac{\chi_\varphi(g)}{d} \overline{\chi_\varphi(g)} = \sum_{i=1}^{s} \sum_{g \in C_i} \left(\frac{1}{d}\chi_i\right) \overline{\chi_i} = \sum_{i=1}^{s} \left(\frac{h_i}{d}\chi_i\right) \overline{\chi_i}. \tag{6.2}$$

But $h_i\chi_i/d$ is an algebraic integer by Theorem 6.2.3, whereas $\overline{\chi_i}$ is an algebraic integer by Corollary 6.2.1 and the closure of algebraic integers under complex conjugation. Since the algebraic integers form a ring, it follows from (6.2) that $|G|/d$ is an algebraic integer, and hence an integer by Proposition 6.1.4. Therefore, $d$ divides $|G|$. □

*Remark 6.2.5.* The dimension theorem was first proved by Frobenius. It was later improved by Schur, who showed that the degree $d$ of an irreducible representation of $G$ divides $[G : Z(G)]$.

The following corollaries are usually proved using facts about $p$-groups and Sylow's theorems.

**Corollary 6.2.6.** *Let $p$ be a prime and let $G$ be a group of order $p^2$. Then $G$ is an abelian.*

*Proof.* Let $d_1, \ldots, d_s$ be the degrees of the irreducible representations of $G$. Then $d_i$ can be $1, p$ or $p^2$. Since the trivial representation has degree 1 and

$$p^2 = |G| = d_1^2 + \cdots + d_s^2$$

it follows that all $d_i = 1$ and hence $G$ is abelian. □

Recall that the *commutator subgroup* $G'$ of a group $G$ is the subgroup generated by all elements of the form $g^{-1}h^{-1}gh$ with $g, h \in G$. It is a normal subgroup and has the properties that $G/G'$ is abelian and if $N$ is any normal subgroup with $G/N$ abelian, then $G' \subseteq N$.

**Lemma 6.2.7.** *Let $G$ be a finite group. Then the number of degree one representations of $G$ divides $|G|$. More precisely, if $G'$ is the commutator subgroup of $G$, then there is a bijection between degree one representations of $G$ and irreducible representations of the abelian group $G/G'$. Hence $G$ has $|G/G'| = [G : G']$ degree one representations.*

*Proof.* Let $\pi \colon G \longrightarrow G/G'$ be the canonical projection $\pi(g) = gG'$. If $\psi \colon G/G' \longrightarrow \mathbb{C}^*$ is an irreducible representation, then $\psi\pi \colon G \longrightarrow \mathbb{C}^*$ is a degree one representation. We now show that every degree one representation of $G$ is obtained in this way. Let $\rho \colon G \longrightarrow \mathbb{C}^*$ be a degree one representation. Then $\operatorname{Im}\rho \cong G/\ker\rho$ is abelian. Therefore, $G' \subseteq \ker\rho$. Define $\psi \colon G/G' \longrightarrow \mathbb{C}^*$ by $\psi(gG') = \rho(g)$. This is well defined because if $gG' = hG'$, then $h^{-1}g \in G' \subseteq \ker\rho$ and so $\rho(h^{-1}g) = 1$. Thus $\rho(h) = \rho(g)$. Clearly $\psi(gG'hG') = \psi(ghG') = \rho(gh) = \rho(g)\rho(h) = \psi(gG')\psi(hG')$ and so $\psi$ is a homomorphism. By construction $\rho = \psi\pi$, completing the proof. □

**Corollary 6.2.8.** *Let $p, q$ be primes with $p < q$ and $q \not\equiv 1 \bmod p$. Then any group $G$ of order $pq$ is abelian.*

*Proof.* Let $d_1, \ldots, d_s$ be the degrees of the irreducible representations of $G$. Since $d_i$ divides $|G|$, $p < q$ and

$$pq = |G| = d_1^2 + \cdots + d_s^2$$

it follows that $d_i = 1, p$ all $i$. Let $n$ be the number of degree $p$ representations of $G$ and let $m$ be the number of degree 1 representations of $G$. Then $pq = m + np^2$. Since $m$ divides $|G|$ by Lemma 6.2.7, $m \geq 1$ (there is at least the trivial representation) and $p \mid m$, we must have $m = p$ or $m = pq$. If $m = p$, then $q = 1 + np$ contradicting that $q \not\equiv 1 \bmod p$. Therefore, $m = pq$ and so all the irreducible representations of $G$ have degree one. Thus $G$ is abelian. □

## 6.3 Burnside's Theorem

Let $G$ be a group of order $n$ and suppose that $\varphi\colon G \longrightarrow GL_d(\mathbb{C})$ is a representation. Then $\chi_\varphi(g)$ is a sum of $d$ $n$th-roots of unity, as was noted in Remark 6.2.2. This explains the relevance of our next lemma.

**Lemma 6.3.1.** *Let $\lambda_1, \ldots, \lambda_d$ be $n$th-roots of unity. Then*

$$|\lambda_1 + \cdots + \lambda_d| \leq d$$

*and equality holds if and only if $\lambda_1 = \lambda_2 = \cdots = \lambda_d$.*

*Proof.* If $v, w \in \mathbb{R}^2$ are vectors, then

$$\|v+w\|^2 = \|v\|^2 + 2\langle v, w\rangle + \|w\|^2 = \|v\|^2 + 2\|v\| \cdot \|w\| \cos\theta + \|w\|^2$$

where $\theta$ is the angle between $v$ and $w$. Since $\cos\theta \leq 1$ with equality if and only if $\theta = 0$, it follows that $\|v+w\| \leq \|v\| + \|w\|$ with equality if and only if $v = \lambda w$ or $w = \lambda v$ some $\lambda \geq 0$. Induction then yields that $\|v_1 + \cdots + v_n\| \leq \|v_1\| + \cdots + \|v_n\|$ with equality if and only if the $v_i$ are non-negative scalar multiples of a common vector.

Identifying complex numbers with vectors in $\mathbb{R}^2$ then yields $|\lambda_1 + \cdots + \lambda_d| \leq |\lambda_1| + \cdots + |\lambda_d|$ with equality if and only if the $\lambda_i$ are non-negative scalar multiples of some complex number $z$. But $|\lambda_1| = \cdots = |\lambda_d| = 1$, so they can only be non-negative multiples of the same complex number if they are the equal. This completes the proof. □

Let $\omega_n = e^{2\pi i/n}$. Denote by $\mathbb{Q}[\omega_n]$ the smallest subfield of $\mathbb{C}$ containing $\omega_n$. This is the smallest subfield $F$ of $\mathbb{C}$ so that $z^n - 1 = (z-\alpha_1)\cdots(z-\alpha_n)$ for some $\alpha_1, \ldots, \alpha_n \in F$, i.e., $F$ is the splitting field of $z^n - 1$. Fields of the form $\mathbb{Q}[\omega_n]$ are called *cyclotomic fields*. Let $\phi$ be the Euler $\phi$-function; so $\phi(n)$ is the number of positive integers less than $n$ that are relatively prime to it. The following is usually proved in a course on rings and fields (cf. [11]).

**Lemma 6.3.2.** *The field $\mathbb{Q}[\omega_n]$ has dimension $\phi(n)$ as a $\mathbb{Q}$-vector space.*

Actually, all we really require is that the dimension is finite, which follows because $\omega_n$ is an algebraic number. We shall also need a little bit of Galois theory. Let $\Gamma = \mathrm{Gal}(\mathbb{Q}[\omega_n] : \mathbb{Q})$ be the Galois group of $\mathbb{Q}[\omega_n]$ over $\mathbb{Q}$. That is, $\Gamma$ is the group of all field automorphisms $\sigma\colon \mathbb{Q}[\omega_n] \longrightarrow \mathbb{Q}[\omega_n]$ such that $\sigma(r) = r$ all $r \in \mathbb{Q}$ (actually this last condition is automatic). It follows from the fundamental theorem of Galois theory that $|\Gamma| = \phi(n)$ since $\dim \mathbb{Q}[\omega_n] = \phi(n)$ as a $\mathbb{Q}$-vector space and $\mathbb{Q}[\omega_n]$ is the splitting field of the polynomial $z^n - 1$. In fact, one can prove that $\Gamma \cong \mathbb{Z}/n\mathbb{Z}^*$, although we shall not use this fact; for us the important thing is that $\Gamma$ is finite.

A crucial fact is that if $p(z)$ is a polynomial with rational coefficients, then $\Gamma$ permutes the roots of $p$ in $\mathbb{Q}[\omega_n]$.

**Lemma 6.3.3.** *Let $p(z)$ be a polynomial with rational coefficients and suppose that $\alpha \in \mathbb{Q}[\omega_n]$ is a root of p. Then $\sigma(\alpha)$ is also a root of p all $\sigma \in \Gamma$.*

*Proof.* Suppose $p(z) = a_k z^k + a_{k-1} z^{k-1} + \cdots + a_0$ with the $a_i \in \mathbb{Q}$. Then

$$\begin{aligned} p(\sigma(\alpha)) &= a_k \sigma(\alpha)^k + a_{k-1}\sigma(\alpha)^{k-1} + \cdots + a_0 \\ &= \sigma(a_k \alpha^k + a_{k-1}\alpha^{k-1} + \cdots + a_0) \\ &= \sigma(0) \\ &= 0 \end{aligned}$$

since $\sigma(a_i) = a_i$ for all $i$. □

**Corollary 6.3.4.** *Let $\alpha$ be an nth-root of unity. Then $\sigma(\alpha)$ is also an nth-root of unity for all $\sigma \in \Gamma$.*

*Proof.* Apply Lemma 6.3.3 to the polynomial $z^n - 1$. □

*Remark 6.3.5.* The proof that $\Gamma \cong \mathbb{Z}/n\mathbb{Z}^*$ follows fairly easily from Corollary 6.3.4; we sketch it here. Since $\Gamma$ permutes the roots of $z^n - 1$, it acts by automorphisms on the cyclic group $C_n = \{\omega_n^k \mid 0 \leq k \leq n-1\}$ of order $n$. As the automorphism group of a cyclic group of order $n$ is isomorphic to $\mathbb{Z}/n\mathbb{Z}^*$, this determines a homomorphism $\tau\colon \Gamma \longrightarrow \mathbb{Z}/n\mathbb{Z}^*$ by $\tau(\sigma) = \sigma|_{C_n}$. Since $\mathbb{Q}[\omega_n]$ is generated over $\mathbb{Q}$ by $\omega_n$, each element of $\Gamma$ is determined by what it does to $\omega_n$ and hence $\tau$ is injective. Since $|\Gamma| = \phi(n) = |\mathbb{Z}/n\mathbb{Z}^*|$, it must be that $\tau$ is an isomorphism.

**Corollary 6.3.6.** *Let $\alpha \in \mathbb{Q}[\omega_n]$ be an algebraic integer and suppose $\sigma \in \Gamma$. Then $\sigma(\alpha)$ is an algebraic integer.*

*Proof.* If $\alpha$ is a root of the monic polynomial $p$ with integer coefficients, then so is $\sigma(\alpha)$ by Lemma 6.3.3. □

Another consequence of the fundamental theorem of Galois theory that we shall require is:

**Theorem 6.3.7.** *Let $\alpha \in \mathbb{Q}[\omega_n]$. Then $\sigma(\alpha) = \alpha$ for all $\sigma \in \Gamma$ if and only if $\alpha \in \mathbb{Q}$.*

The following corollary is a thinly veiled version of our old friend, the averaging trick.

**Corollary 6.3.8.** *Let $\alpha \in \mathbb{Q}[\omega_n]$. Then $\prod_{\sigma \in \Gamma} \sigma(\alpha) \in \mathbb{Q}$.*

*Proof.* Let $\tau \in \Gamma$. Then we have

$$\tau\left(\prod_{\sigma \in \Gamma} \sigma(\alpha)\right) = \prod_{\sigma \in \Gamma} \tau\sigma(\alpha) = \prod_{\rho \in \Gamma} \rho(\alpha)$$

where the last equality is obtained by setting $\rho = \tau\sigma$. Theorem 6.3.7 now yields the desired conclusion. □

The next theorem is of a somewhat technical nature (meaning that I do not see how to motivate it), but is crucial to proving Burnside's theorem.

**Theorem 6.3.9.** *Let $G$ be a group of order $n$ and let $C$ be a conjugacy class of $G$. Suppose that $\varphi\colon G \longrightarrow GL_d(\mathbb{C})$ is an irreducible representation and assume that $h = |C|$ is relatively prime to $d$. Then either:*

1. *there exists $\lambda \in \mathbb{C}^*$ such that $\varphi_g = \lambda I$ for all $g \in C$; or*
2. *$\chi_\varphi(g) = 0$ for all $g \in C$.*

*Proof.* Set $\chi = \chi_\varphi$. First note that if $\varphi_g = \lambda I$ for some $g \in C$, then $\varphi_x = \lambda I$ for all $x \in C$ since conjugating a scalar matrix does not change it. Also since $\chi$ is a class function, if it vanishes on any element of $C$, it must vanish on all elements of $C$. Therefore, it suffices to show that if $\varphi_g \neq \lambda I$ for some $g \in C$, then $\chi_\varphi(g) = 0$.

By Theorem 6.2.3 we know that $h\chi(g)/d$ is an algebraic integer; also $\chi(g)$ is an algebraic integer by Corollary 6.2.1. Since $\gcd(d, h) = 1$, we can find integers $k, j$ so that $kh + jd = 1$. Let

$$\alpha = k\left(\frac{h}{d}\chi(g)\right) + j\chi(g) = \frac{kh + jd}{d}\chi(g) = \frac{\chi(g)}{d}.$$

Then $\alpha$ is an algebraic integer. By Corollary 4.1.10, $\varphi_g$ is diagonalizable and its eigenvalues $\lambda_1, \ldots, \lambda_d$ are $n$th-roots of unity. Since $\varphi_g$ is diagonalizable but not a scalar matrix, its eigenvalues are not all the same. Applying Lemma 6.3.1 to $\chi(g) = \lambda_1 + \cdots + \lambda_d$ yields $|\chi(g)| < d$, and so

$$|\alpha| = \left|\frac{\chi(g)}{d}\right| < 1.$$

Also note that $\alpha \in \mathbb{Q}[\omega_n]$. Let $\sigma \in \Gamma$. Lemma 6.3.6 implies that $\sigma(\alpha)$ is an algebraic integer. Corollary 6.3.4 tells us that

$$\sigma(\chi(g)) = \sigma(\lambda_1) + \cdots + \sigma(\lambda_d)$$

is again a sum of $d$ $nth$-roots of unity, not all equal. Hence, another application of Lemma 6.3.1 yields

$$|\sigma(\alpha)| = \left|\frac{\sigma(\chi(g))}{d}\right| < 1.$$

Putting all this together, we obtain that $q = \prod_{\sigma\in\Gamma} \sigma(\alpha)$ is an algebraic integer with

$$|q| = \left|\prod_{\sigma\in\Gamma} \sigma(\alpha)\right| = \prod_{\sigma\in\Gamma} |\sigma(\alpha)| < 1.$$

But Corollary 6.3.8 tells us that $q \in \mathbb{Q}$. Therefore, $q \in \mathbb{Z}$ by Proposition 6.1.4. Since $|q| < 1$, we may conclude that $q = 0$ and hence $\sigma(\alpha) = 0$ for some $\sigma \in \Gamma$. Because

$\sigma$ is an automorphism, this implies $\alpha = 0$. We conclude $\chi(g) = 0$, as was to be proved. □

We are just one lemma away from proving Burnside's theorem.

**Lemma 6.3.10.** *Let $G$ be a finite non-abelian group. Suppose that there is a conjugacy class $C \neq \{1\}$ of $G$ such that $|C| = p^t$ with $p$ prime, $t \geq 0$. Then $G$ is not simple.*

*Proof.* Assume that $G$ is simple and let $\varphi^{(1)}, \ldots, \varphi^{(s)}$ be a complete set of representatives of the equivalence classes of irreducible representations of $G$. Let $\chi_1, \ldots, \chi_s$ be their respective characters and $d_1, \ldots, d_s$ their degrees. We may take $\varphi^{(1)}$ to be the trivial representation. Since $G$ is simple, $\ker \varphi^{(k)} = \{1\}$ for $k > 1$ (since $\ker \varphi^{(k)} = G$ implies $\varphi^{(k)}$ is the trivial representation). Therefore, $\varphi^{(k)}$ is injective for $k > 1$ and so, since $G$ is non-abelian and $\mathbb{C}^*$ is abelian, it follows that $d_k > 1$ for $k > 1$. Also, because $G$ is simple, non-abelian $Z(G) = \{1\}$ and so $t > 0$.

Let $g \in C$ and $k > 1$. Let $Z_k$ be the set of all elements $x \in G$ such that $\varphi_x^{(k)}$ is a scalar matrix. Let $H = \{\lambda I_{d_k} \mid \lambda \in \mathbb{C}^*\}$; then $H$ is a subgroup of $GL_{d_k}(\mathbb{C})$ contained in the center, and hence normal (actually it is the center). As $Z_k$ is the inverse image of $H$ under $\varphi^{(k)}$, we conclude that $Z_k$ is a normal subgroup of $G$. Since $d_k > 1$, we cannot have $Z_k = G$. Thus $Z_k = \{1\}$ by simplicity of $G$. Suppose for the moment that $p \nmid d_k$; then $\chi_k(g) = 0$ by Theorem 6.3.9.

Let $L$ be the regular representation of $G$. Recall $L \sim d_1\varphi^{(1)} \oplus \cdots \oplus d_s\varphi^{(s)}$. Since $g \neq 1$, Proposition 4.4.3 yields

$$\begin{aligned}
0 = \chi_L(g) &= d_1\chi_1(g) + \cdots + d_s\chi_s(g) \\
&= 1 + \sum_{k=2}^{s} d_k\chi_k(g) \\
&= 1 + \sum_{p \mid d_k} d_k\chi_k(g) \\
&= 1 + pz
\end{aligned}$$

where $z$ is an algebraic integer. Hence $1/p = -z$ is an algebraic integer, and thus an integer by Proposition 6.1.4. This contradiction establishes the lemma. □

**Theorem 6.3.11 (Burnside).** *Let $G$ be a group of order $p^a q^b$ with $p, q$ primes. Then $G$ is not simple unless it is cyclic of prime order.*

*Proof.* Since an abelian group is simple if and only if it is cyclic of prime order, we may assume that $G$ is non-abelian. Since groups of prime power order have non-trivial centers [11], if $a$ or $b$ is zero, then we are done. Suppose next that $a, b \geq 1$. By Sylow's theorem [11], $G$ has a subgroup $H$ of order $q^b$. Let $1 \neq g \in Z(H)$ and let $N_G(g) = \{x \in G \mid xg = gx\}$ be the normalizer of $g$ in $G$. Then $H \subseteq N_G(g)$ as $g \in Z(H)$. Thus

$$p^a = [G : H] = [G : N_G(g)][N_G(g) : H]$$

and so $[G : N_G(g)] = p^t$ for some $t \geq 0$. But $[G : N_G(g)]$ is the size of the conjugacy class of $g$. The previous lemma now implies $G$ is not simple. □

*Remark 6.3.12.* Burnside's theorem is often stated in the equivalent form that all groups of order $p^a q^b$, with $p, q$ primes, are solvable.

## Exercises

**Exercise 6.1.** Let $F$ be a subfield of $\mathbb{C}$. Prove that every automorphism of $F$ fixes $\mathbb{Q}$.

**Exercise 6.2.** Let $G$ be a non-abelian group of order 39.

1. Determine the degrees of the irreducible representations of $G$ and how many irreducible representations $G$ has of each degree (up to equivalence).
2. Determine the number of conjugacy classes of $G$.

**Exercise 6.3.** Let $G$ be a non-abelian group of order 21.

1. Determine the degrees of the irreducible representations of $G$ and how many irreducible representations $G$ has of each degree (up to equivalence).
2. Determine the number of conjugacy classes of $G$.

**Exercise 6.4.** Prove that if there is a non-solvable group of order $p^a q^b$ with $p, q$ primes, then there is a simple non-abelian group of order $p^{a'} q^{b'}$.

**Exercise 6.5.** Show that if $\varphi\colon G \longrightarrow GL_d(\mathbb{C})$ is a representation with character $\chi$, then $g \in \ker \varphi$ if and only if $\chi(g) = d$. (Hint: Use Corollary 4.1.10 and Lemma 6.3.1.)

**Exercise 6.6.** For $\alpha \in \mathbb{C}$, denote by $\mathbb{Z}[\alpha]$ the smallest subring of $\mathbb{C}$ containing $\alpha$.

1. Prove that $\mathbb{Z}[\alpha] = \{a_0 + a_1\alpha + \cdots + a_n\alpha^n \mid n \in \mathbb{N}, a_0, \ldots, a_n \in \mathbb{Z}\}$.
2. Prove that the following are equivalent:

   (a) $\alpha$ is an algebraic integer;
   (b) The additive group of $\mathbb{Z}[\alpha]$ is finitely generated;
   (c) $\alpha$ is contained in a finitely generated subgroup of the additive group of $\mathbb{C}$, which is closed under multiplication by $\alpha$.

   (Hint: for (c) implies (a) use Lemma 6.1.5.)

# Chapter 7
# Group Actions and Permutation Representations

In this chapter we link representation theory with the theory of group actions and permutation groups. Once again, we are only able to provide a brief glimpse of these connections; see [3] for more. In this chapter all groups are assumed to be finite and all actions of groups are taken to be on finite sets.

## 7.1 Group Actions

Let us begin by recalling the definition of a group action. If $X$ is a set, then $S_X$ will denote the symmetric group on $X$. We shall tacitly assume $|X| \geq 2$, as the case $|X| = 1$ is uninteresting.

**Definition 7.1.1 (Group action).** An *action* of a group $G$ on a set $X$ is a homomorphism $\sigma\colon G \longrightarrow S_X$. We often write $\sigma_g$ for $\sigma(g)$. The cardinality of $X$ is called the *degree* of the action.

*Example 7.1.2 (Regular action).* Let $G$ be a group and define $\lambda\colon G \longrightarrow S_G$ by $\lambda_g(x) = gx$. Then $\lambda$ is called the *regular action* of $G$ on $G$.

A subset $Y \subseteq X$ is called *$G$-invariant* if $\sigma_g(y) \in Y$ for all $y \in Y$, $g \in G$. One can always partition $X$ into a disjoint union of minimal $G$-invariant subsets called orbits.

**Definition 7.1.3 (Orbit).** Let $\sigma\colon G \longrightarrow S_X$ be a group action. The *orbit* of $x \in X$ under $G$ is the set $G \cdot x = \{\sigma_g(x) \mid g \in G\}$.

Clearly, the orbits are $G$-invariant. A standard course in group theory proves that distinct orbits are disjoint and the union of all the orbits is $X$, that is, the orbits form a partition of $X$ (cf. [11]). Of particular importance is the case when there is just one orbit.

B. Steinberg, *Representation Theory of Finite Groups: An Introductory Approach*, Universitext, DOI 10.1007/978-1-4614-0776-8_7,
© Springer Science+Business Media, LLC 2012

**Definition 7.1.4 (Transitive).** A group action $\sigma\colon G \longrightarrow S_X$ is *transitive* if, for all $x, y \in X$, there exists $g \in G$ such that $\sigma_g(x) = y$. Equivalently, the action is transitive if there is one orbit of $G$ on $X$.

*Example 7.1.5 (Coset action).* If $G$ is a group and $H$ a subgroup, then there is an action $\sigma\colon G \longrightarrow S_{G/H}$ given by $\sigma_g(xH) = gxH$. This action is transitive.

An even stronger property than transitivity is that of 2-transitivity.

**Definition 7.1.6 (2-transitive).** An action $\sigma\colon G \longrightarrow S_X$ of $G$ on $X$ is *2-transitive* if given any two pairs of distinct elements $x, y \in X$ and $x', y' \in X$, there exists $g \in G$ such that $\sigma_g(x) = x'$ and $\sigma_g(y) = y'$.

*Example 7.1.7 (Symmetric groups).* For $n \geq 2$, the action of $S_n$ on $\{1, \ldots, n\}$ is 2-transitive. Indeed, let $i \neq j$ and $k \neq \ell$ be pairs of elements of $X$. Let $X = \{1, \ldots, n\} \setminus \{i, j\}$ and $Y = \{1, \ldots, n\} \setminus \{k, \ell\}$. Then $|X| = n - 2 = |Y|$, so we can choose a bijection $\alpha\colon X \longrightarrow Y$. Define $\tau \in S_n$ by

$$\tau(m) = \begin{cases} k & m = i \\ \ell & m = j \\ \alpha(m) & \text{else.} \end{cases}$$

Then $\tau(i) = k$ and $\tau(j) = \ell$. This establishes that $S_n$ is 2-transitive.

Let us put this notion into a more general context.

**Definition 7.1.8 (Orbital).** Let $\sigma\colon G \longrightarrow S_X$ be a transitive group action. Define $\sigma^2\colon G \longrightarrow S_{X \times X}$ by

$$\sigma_g^2(x_1, x_2) = (\sigma_g(x_1), \sigma_g(x_2)).$$

An orbit of $\sigma^2$ is termed an *orbital* of $\sigma$. The number of orbitals is called the *rank* of $\sigma$.

Let $\Delta = \{(x, x) \mid x \in X\}$. As $\sigma_g^2(x, x) = (\sigma_g(x), \sigma_g(x))$, it follows from the transitivity of $G$ on $X$ that $\Delta$ is an orbital. It is called the *diagonal* or *trivial orbital*.

*Remark 7.1.9.* Orbitals are closely related to graph theory. If $G$ acts transitively on $X$, then any non-trivial orbital can be viewed as the edge set of a graph with vertex set $X$ (by symmetrizing). The group $G$ acts on the resulting graph as a vertex-transitive group of automorphisms.

**Proposition 7.1.10.** *Let $\sigma\colon G \longrightarrow S_X$ be a group action (with $X \geq 2$). Then $\sigma$ is 2-transitive if and only if $\sigma$ is transitive and* $\operatorname{rank}(\sigma) = 2$.

*Proof.* First we observe that transitivity is necessary for 2-transitivity since if $G$ is 2-transitive on $X$ and $x, y \in X$, then we may choose $x' \neq x$ and $y' \neq y$.

By 2-transitivity there exists $g \in G$ with $\sigma_g(x) = y$ and $\sigma_g(x') = y'$. This shows that $\sigma$ is transitive. Next observe that

$$(X \times X) \setminus \Delta = \{(x, y) \mid x \neq y\}$$

and so the complement of $\Delta$ is an orbital if and only if, for any two pairs $x \neq y$ and $x' \neq y'$ of distinct elements, there exists $g \in G$ with $\sigma_g(x) = x'$ and $\sigma_g(y) = y'$, that is, $\sigma$ is 2-transitive. □

Consequently the rank of $S_n$ is 2. Let $\sigma\colon G \longrightarrow S_X$ be a group action. Then, for $g \in G$, we define

$$\mathrm{Fix}(g) = \{x \in X \mid \sigma_g(x) = x\}$$

to be the set of *fixed points* of $g$. Let $\mathrm{Fix}^2(g)$ be the set of fixed points of $g$ on $X \times X$. The notation is unambiguous because of the following proposition.

**Proposition 7.1.11.** *Let $\sigma\colon G \longrightarrow S_X$ be a group action. Then the equality*

$$\mathrm{Fix}^2(g) = \mathrm{Fix}(g) \times \mathrm{Fix}(g)$$

*holds. Hence $|\mathrm{Fix}^2(g)| = |\mathrm{Fix}(g)|^2$.*

*Proof.* Let $(x, y) \in X \times X$. Then $\sigma_g^2(x, y) = (\sigma_g(x), \sigma_g(y))$ and so $(x, y) = \sigma_g^2(x, y)$ if and only if $\sigma_g(x) = x$ and $\sigma_g(y) = y$. We conclude $\mathrm{Fix}^2(g) = \mathrm{Fix}(g) \times \mathrm{Fix}(g)$. □

## 7.2 Permutation Representations

Given a group action $\sigma\colon G \longrightarrow S_n$, we may compose it with the standard representation $\alpha\colon S_n \longrightarrow GL_n(\mathbb{C})$ to obtain a representation of $G$. Let us formalize this.

**Definition 7.2.1 (Permutation representation).** Let $\sigma\colon G \longrightarrow S_X$ be a group action. Define a representation $\widetilde{\sigma}\colon G \longrightarrow GL(\mathbb{C}X)$ by setting

$$\widetilde{\sigma}_g\left(\sum_{x \in X} c_x x\right) = \sum_{x \in X} c_x \sigma_g(x) = \sum_{y \in X} c_{\sigma_{g^{-1}}(y)} y.$$

One calls $\widetilde{\sigma}$ the *permutation representation* associated to $\sigma$.

*Remark 7.2.2.* Notice that $\widetilde{\sigma}_g$ is the linear extension of the map defined on the basis $X$ of $\mathbb{C}X$ by sending $x$ to $\sigma_g(x)$. Also, observe that the degree of the representation $\widetilde{\sigma}$ is the same as the degree of the group action $\sigma$.

*Example 7.2.3 (Regular representation).* Let $\lambda\colon G \longrightarrow S_G$ be the regular action. Then one has $\widetilde{\lambda} = L$, the regular representation.

The following proposition is proved exactly as in the case of the regular representation (cf. Proposition 4.4.2), so we omit the proof.

**Proposition 7.2.4.** *Let $\sigma\colon G \longrightarrow S_X$ be a group action. Then the permutation representation $\widetilde{\sigma}\colon G \longrightarrow GL(\mathbb{C}X)$ is a unitary representation of $G$.*

Next we compute the character of $\widetilde{\sigma}$.

**Proposition 7.2.5.** *Let $\sigma\colon G \longrightarrow S_X$ be a group action. Then*

$$\chi_{\widetilde{\sigma}}(g) = |\mathrm{Fix}(g)|.$$

*Proof.* Let $X = \{x_1, \ldots, x_n\}$ and let $[\widetilde{\sigma}_g]$ be the matrix of $\widetilde{\sigma}$ with respect to this basis. Then $\widetilde{\sigma}_g(x_j) = \sigma_g(x_j)$, so

$$[\widetilde{\sigma}_g]_{ij} = \begin{cases} 1 & x_i = \sigma_g(x_j) \\ 0 & \text{else.} \end{cases}$$

In particular,

$$\begin{aligned} [\widetilde{\sigma}_g]_{ii} &= \begin{cases} 1 & x_i = \sigma_g(x_i) \\ 0 & \text{else} \end{cases} \\ &= \begin{cases} 1 & x_i \in \mathrm{Fix}(g) \\ 0 & \text{else} \end{cases} \end{aligned}$$

and so $\chi_{\widetilde{\sigma}}(g) = \mathrm{Tr}([\widetilde{\sigma}_g]) = |\mathrm{Fix}(g)|$. □

Like the regular representation, permutation representations are never irreducible (if $|X| > 1$). To understand better how it decomposes, we first consider the trivial component.

**Definition 7.2.6 (Fixed subspace).** Let $\varphi\colon G \longrightarrow GL(V)$ be a representation. Then

$$V^G = \{v \in V \mid \varphi_g(v) = v \text{ for all } g \in G\}$$

is the *fixed subspace* of $G$.

One easily verifies that $V^G$ is a $G$-invariant subspace and the subrepresentation $\varphi|_{V^G}$ is equivalent to $\dim V^G$ copies of the trivial representation. Let us prove that $V^G$ is the direct sum of all the copies of the trivial representation in $\varphi$.

**Proposition 7.2.7.** *Let $\varphi\colon G \longrightarrow GL(V)$ be a representation and let $\chi_1$ be the trivial character of $G$. Then $\langle \chi_\varphi, \chi_1 \rangle = \dim V^G$.*

*Proof.* Write $V = m_1V_1 \oplus \cdots \oplus m_sV_s$ where $V_1, \ldots, V_s$ are irreducible $G$-invariant subspaces whose associated subrepresentations range over the distinct equivalence classes of irreducible representations of $G$ (we allow $m_i = 0$). Without loss of generality, we may assume that $V_1$ is equivalent to the trivial representation. Let $\varphi^{(i)}$ be the restriction of $\varphi$ to $V_i$. Now if $v \in V$, then $v = v_1 + \cdots + v_s$ with the $v_i \in m_i V_i$ and

$$\varphi_g v = (m_1\varphi^{(1)})_g v_1 + \cdots + (m_s\varphi^{(s)})_g v_s = v_1 + (m_2\varphi^{(2)})_g v_2 + \cdots + (m_s\varphi^{(s)})_g v_s$$

and so $g \in V^G$ if and only if $v_i \in m_i V_i^G$ for all $2 \leq i \leq s$. In other words,

$$V^G = m_1 V_1 \oplus m_2 V_2^G \oplus \cdots \oplus m_s V_s^G.$$

Let $i \geq 2$. Since $V_i$ is irreducible and not equivalent to the trivial representation and $V_i^G$ is $G$-invariant, it follows $V_i^G = 0$. Thus $V^G = m_1 V_1$ and so the multiplicity of the trivial representation in $\varphi$ is $\dim V^G$, as required. □

Now we compute $\mathbb{C}X^G$ when we have a permutation representation.

**Proposition 7.2.8.** *Let* $\sigma\colon G \longrightarrow S_X$ *be a group action. Let* $\mathcal{O}_1, \ldots, \mathcal{O}_m$ *be the orbits of* $G$ *on* $X$ *and define* $v_i = \sum_{x \in \mathcal{O}_i} x$. *Then* $v_1, \ldots, v_m$ *is a basis for* $\mathbb{C}X^G$ *and hence* $\dim \mathbb{C}X^G$ *is the number of orbits of* $G$ *on* $X$.

*Proof.* First observe that

$$\widetilde{\sigma}_g v_i = \sum_{x \in \mathcal{O}_i} \sigma_g(x) = \sum_{y \in \mathcal{O}_i} y = v_i$$

as is seen by setting $y = \sigma_g(x)$ and using that $\sigma_g$ permutes the orbit $\mathcal{O}_i$. Thus $v_1, \ldots, v_m \in \mathbb{C}X^G$. Since the orbits are disjoint, we have

$$\langle v_i, v_j \rangle = \begin{cases} |\mathcal{O}_i| & i = j \\ 0 & i \neq j \end{cases}$$

and so $\{v_1, \ldots, v_s\}$ is an orthogonal set of non-zero vectors and hence linearly independent. It remain to prove that this set spans $\mathbb{C}X^G$.

Suppose $v = \sum_{x \in X} c_x x \in \mathbb{C}X^G$. We show that if $z \in G \cdot y$, then $c_y = c_z$. Indeed, let $z = \sigma_g(y)$. Then we have

$$\sum_{x \in X} c_x x = v = \widetilde{\sigma}_g v = \sum_{x \in X} c_x \sigma_g(x) \tag{7.1}$$

and so the coefficient of $z$ in the left-hand side of (7.1) is $c_z$, whereas the coefficient of $z$ in the right-hand side is $c_y$ since $z = \sigma_g(y)$. Thus $c_z = c_y$. It follows that there are complex numbers $c_1, \ldots, c_m$ such that $c_x = c_i$ for all $x \in \mathcal{O}_i$. Thus

$$v = \sum_{x \in X} c_x x = \sum_{i=1}^{m} \sum_{x \in \mathcal{O}_i} c_x x = \sum_{i=1}^{m} c_i \sum_{x \in \mathcal{O}_i} x = \sum_{i=1}^{m} c_i v_i$$

and hence $v_1, \ldots, v_m$ span $\mathbb{C}X^G$, completing the proof. □

Since $G$ always has at least one orbit on $X$, the above result shows that the trivial representation appears as a constituent in $\widetilde{\sigma}$ and so if $|X| > 1$, then $\widetilde{\sigma}$ is not irreducible. As a corollary to the above proposition we prove a useful result, commonly known as *Burnside's lemma*, although it seems to have been known earlier to Cauchy and Frobenius. It has many applications in combinatorics to counting problems. The lemma says that the number of orbits of $G$ on $X$ is the average number of fixed points.

**Corollary 7.2.9 (Burnside's lemma).** *Let $\sigma\colon G \longrightarrow S_X$ be a group action and let $m$ be the number of orbits of $G$ on $X$. Then*

$$m = \frac{1}{|G|}\sum_{g\in G} |\mathrm{Fix}(g)|.$$

*Proof.* Let $\chi_1$ be the trivial character of $G$. By Propositions 7.2.5, 7.2.7 and 7.2.8 we have

$$m = \langle \chi_{\widetilde{\sigma}}, \chi_1\rangle = \frac{1}{|G|}\sum_{g\in G} \chi_{\widetilde{\sigma}}(g)\overline{\chi_1(g)} = \frac{1}{|G|}\sum_{g\in G} |\mathrm{Fix}(g)|$$

as required. □

As a corollary, we obtain two formulas for the rank of $\sigma$.

**Corollary 7.2.10.** *Let $\sigma\colon G \longrightarrow S_X$ be a transitive group action. Then the equalities*

$$\mathrm{rank}(\sigma) = \frac{1}{|G|}\sum_{g\in G} |\mathrm{Fix}(g)|^2 = \langle \chi_{\widetilde{\sigma}}, \chi_{\widetilde{\sigma}}\rangle$$

*hold.*

*Proof.* Since $\mathrm{rank}(\sigma)$ is the number of orbits of $\sigma^2$ on $X \times X$ and the number of fixed points of $g$ on $X \times X$ is $|\mathrm{Fix}(g)|^2$ (Proposition 7.1.11), the first equality is a consequence of Burnside's lemma. For the second, we compute

$$\langle \chi_{\widetilde{\sigma}}, \chi_{\widetilde{\sigma}}\rangle = \frac{1}{|G|}\sum_{g\in G} |\mathrm{Fix}(g)|\overline{|\mathrm{Fix}(g)|} = \frac{1}{|G|}\sum_{g\in G} |\mathrm{Fix}(g)|^2$$

completing the proof. □

Assume now that $\sigma\colon G \longrightarrow S_X$ is a transitive action. Let $v_0 = \sum_{x\in X} x$. Then $\mathbb{C}X^G = \mathbb{C}v_0$ by Proposition 7.2.8. Since $\widetilde{\sigma}$ is a unitary representation, $V_0 = \mathbb{C}v_0^{\perp}$ is a $G$-invariant subspace (cf. the proof of Proposition 3.2.3). Usually, $\mathbb{C}v_0$ is called the *trace* of $\sigma$ and $V_0$ the *augmentation* of $\sigma$. Let $\widetilde{\sigma}'$ be the restriction of $\widetilde{\sigma}$ to $V_0$; we call it the *augmentation representation* associated to $\sigma$. As $\mathbb{C}X = V_0 \oplus \mathbb{C}v_0$, it follows that $\chi_{\widetilde{\sigma}} = \chi_{\widetilde{\sigma}'} + \chi_1$ where $\chi_1$ is the trivial character. We now characterize when the augmentation representation $\widetilde{\sigma}'$ is irreducible.

**Theorem 7.2.11.** *Let $\sigma\colon G \longrightarrow S_X$ be a transitive group action. Then the augmentation representation $\widetilde{\sigma}'$ is irreducible if and only if $G$ is 2-transitive on $X$.*

*Proof.* This is a simple calculation using Corollary 7.2.10 and the fact that $G$ is 2-transitive on $X$ if and only if $\mathrm{rank}(\sigma) = 2$ (Proposition 7.1.10). Indeed, if $\chi_1$ is the trivial character of $G$, then

$$\begin{aligned}\langle \chi_{\widetilde{\sigma}'}, \chi_{\widetilde{\sigma}'}\rangle &= \langle \chi_{\widetilde{\sigma}} - \chi_1, \chi_{\widetilde{\sigma}} - \chi_1\rangle \\ &= \langle \chi_{\widetilde{\sigma}}, \chi_{\widetilde{\sigma}}\rangle - \langle \chi_{\widetilde{\sigma}}, \chi_1\rangle - \langle \chi_1, \chi_{\widetilde{\sigma}}\rangle + \langle \chi_1, \chi_1\rangle. \end{aligned} \quad (7.2)$$

Now by Proposition 7.2.8 $\langle \chi_{\widetilde{\sigma}}, \chi_1\rangle = 1$ because $G$ is transitive. Therefore, $\langle \chi_1, \chi_{\widetilde{\sigma}}\rangle = 1$. Also, $\langle \chi_1, \chi_1\rangle = 1$. Thus (7.2) becomes, in light of Corollary 7.2.10,

$$\langle \chi_{\widetilde{\sigma}'}, \chi_{\widetilde{\sigma}'}\rangle = \mathrm{rank}(\sigma) - 1$$

and so $\chi_{\widetilde{\sigma}'}$ is an irreducible character if and only if $\mathrm{rank}(\sigma) = 2$, that is, if and only if $G$ is 2-transitive on $X$. □

*Remark 7.2.12.* The decomposition of the standard representation of $S_3$ in Example 4.3.17 corresponds precisely to the decomposition into the direct sum of the augmentation and the trace.

With Theorem 7.2.11 in hand, we may now compute the character table of $S_4$.

*Example 7.2.13 (Character table of $S_4$).* First of all $S_4$ has five conjugacy classes, represented by $Id, (1\ 2), (1\ 2\ 3), (1\ 2\ 3\ 4), (1\ 2)(3\ 4)$. Let $\chi_1$ be the trivial character and $\chi_2$ the character of the sign homomorphism. As $S_4$ acts 2-transitively on $\{1,\ldots,4\}$, Theorem 7.1.10 implies that the augmentation representation is irreducible. Let $\chi_4$ be the character of this representation; it is the character of the standard representation minus the trivial character so $\chi_4(g) = |\mathrm{Fix}(g)| - 1$. Let $\chi_5 = \chi_2 \cdot \chi_4$. That is, if $\tau$ is the representation associated to $\chi_4$, then we can define a new representation $\tau^{\chi_2}\colon S_4 \longrightarrow GL_3(\mathbb{C})$ by $\tau^{\chi_2}_g = \chi_2(g)\tau_g$. It is easily verified that $\chi_{\tau^{\chi_2}}(g) = \chi_2(g)\chi_4(g)$ and $\tau^{\chi_2}$ is irreducible. This gives us four of the five irreducible representations. How do we get the fifth? Let $d$ be the degree of the missing representation. Then

$$24 = |S_4| = 1^2 + 1^2 + d^2 + 3^2 + 3^2 = 20 + d^2$$

and so $d = 2$. Let $\chi_3$ be the character of the missing irreducible representation and let $L$ be the regular representation of $S_4$. Then

$$\chi_L = \chi_1 + \chi_2 + 2\chi_3 + 3\chi_4 + 3\chi_5$$

so for $Id \neq g \in S_4$, we have

$$\chi_3(g) = \frac{1}{2}\left(-\chi_1(g) - \chi_2(g) - 3\chi_4(g) - 3\chi_5(g)\right).$$

In this way we are able to produce the character table of $S_4$ in Table 7.1.

**Table 7.1** Character table of $S_4$

| | Id | (1 2) | (1 2 3) | (1 2 3 4) | (1 2)(3 4) |
|---|---|---|---|---|---|
| $\chi_1$ | 1 | 1 | 1 | 1 | 1 |
| $\chi_2$ | 1 | −1 | 1 | −1 | 1 |
| $\chi_3$ | 2 | 0 | −1 | 0 | 2 |
| $\chi_4$ | 3 | 1 | 0 | −1 | −1 |
| $\chi_5$ | 3 | −1 | 0 | 1 | −1 |

The reader should try to produce a representation with character $\chi_3$. As a hint, observe that $K = \{Id, (1\ 2)(3\ 4), (1\ 3)(2\ 4), (1\ 4)(2\ 3)\}$ is a normal subgroup of $S_4$ and that $S_4/K \cong S_3$. Construct an irreducible representation by composing the surjective map $S_4 \longrightarrow S_3$ with the degree 2 irreducible representation of $S_3$ coming from the augmentation representation for $S_3$.

## 7.3 The Centralizer Algebra and Gelfand Pairs

Let $\sigma\colon G \longrightarrow S_X$ be a transitive group action. Our goal in this section is to study the ring $\mathrm{Hom}_G(\widetilde{\sigma}, \widetilde{\sigma})$. We only scratch the surface of this topic in this section. Much more information, as well as applications to probability and statistics, can be found in [3].

Let us assume that $X = \{x_1, \ldots, x_n\}$. Define a matrix representation $\varphi\colon G \longrightarrow GL_n(\mathbb{C})$ by $\varphi_g = [\widetilde{\sigma}_g]_X$. Then $\varphi \sim \widetilde{\sigma}$ and so $\mathrm{Hom}_G(\widetilde{\sigma}, \widetilde{\sigma}) \cong \mathrm{Hom}_G(\varphi_g, \varphi_g)$. Next observe that

$$\begin{aligned}\mathrm{Hom}_G(\varphi, \varphi) &= \{A \in M_n(\mathbb{C}) \mid A\varphi_g = \varphi_g A, \forall g \in G\}\\ &= \{A \in M_n(\mathbb{C}) \mid \varphi_g A \varphi_g^{-1} = A, \forall g \in G\}.\end{aligned}$$

From now on we denote $\mathrm{Hom}_G(\varphi, \varphi)$ by $C(\sigma)$ and call it the *centralizer algebra* of $\sigma$.

**Proposition 7.3.1.** *$C(\sigma)$ is a unital subring of $M_n(\mathbb{C})$.*

*Proof.* Trivially, $\varphi_g I_n \varphi_g^{-1} = I_n$ for all $g \in G$. If $A, B \in C(\sigma)$, then

$$\varphi_g(A + B)\varphi_g^{-1} = \varphi_g A \varphi_g^{-1} + \varphi_g B \varphi_g^{-1} = A + B$$

for all $g \in G$, and similarly $\varphi_g(AB)\varphi_g^{-1} = \varphi_g A \varphi_g^{-1} \varphi_g B \varphi_g^{-1} = AB$. Thus $C(\sigma)$ is indeed a unital subring of $M_n(\mathbb{C})$. □

We aim to show that $\dim C(\sigma) = \mathrm{rank}(\sigma)$ and exhibit an explicit basis. Let $V = M_n(\mathbb{C})$ and define a representation $\tau\colon G \longrightarrow GL(V)$ by $\tau_g(A) = \varphi_g A \varphi_g^{-1}$. The reader should perform the routine verification that $\tau$ is indeed a representation. Notice that

$$V^G = \{A \in M_n(\mathbb{C}) \mid \varphi_g A \varphi_g^{-1} = A, \forall g \in G\} = C(\sigma).$$

Let $\sigma^2 \colon G \longrightarrow S_{X\times X}$ be as per Definition 7.1.8. We exhibit an explicit equivalence between $\tau$ and $\widetilde{\sigma^2}$. We can then use Proposition 7.2.8 to obtain a basis for $C(\sigma)$.

**Proposition 7.3.2.** *Define a mapping* $T \colon M_n(\mathbb{C}) \longrightarrow \mathbb{C}(X \times X)$ *by*

$$T(a_{ij}) = \sum_{i,j=1}^{n} a_{ij}(x_i, x_j)$$

*where we have retained the above notation. Then* $T$ *is an equivalence between* $\tau$ *and* $\widetilde{\sigma^2}$.

*Proof.* The map $T$ is evidently bijective and linear with inverse

$$\sum_{i,j=1}^{n} a_{ij}(x_i, x_j) \longmapsto (a_{ij}).$$

Let us check that it is an equivalence. Let $g \in G$ and let $A = (a_{ij}) \in M_n(\mathbb{C})$. Put $B = \tau_g A$; say $B = (b_{ij})$. Define an action $\gamma \colon G \longrightarrow S_n$ by $\sigma_g(x_i) = x_{\gamma_g(i)}$ for $g \in G$. Then

$$b_{ij} = \sum_{k=1,\ell=1}^{n} \varphi(g)_{ik} a_{k\ell} \varphi(g^{-1})_{\ell j} = a_{\gamma_g^{-1}(i),\gamma_g^{-1}(j)}$$

because

$$\varphi(g)_{ik} = \begin{cases} 1 & x_i = \sigma_g(x_k) \\ 0 & \text{else} \end{cases} \quad \text{and} \quad \varphi(g^{-1})_{\ell j} = \begin{cases} 1 & x_\ell = \sigma_g^{-1}(x_j) \\ 0 & \text{else.} \end{cases}$$

Therefore, we have

$$\begin{aligned} T\tau_g A &= \sum_{i,j=1}^{n} b_{ij}(x_i, x_j) = \sum_{i,j=1}^{n} a_{\gamma_g^{-1}(i),\gamma_g^{-1}(j)}(x_i, x_j) \\ &= \sum_{i,j=1}^{n} a_{ij}(\sigma_g(x_i), \sigma_g(x_j)) = \sum_{i,j=1}^{n} a_{ij}\sigma_g^2(x_i, x_j) = \widetilde{\sigma^2}_g TA \end{aligned}$$

and so $T$ is an equivalence, as required. □

We can now provide a basis for $C(\sigma)$. If $\Omega$ is an orbital of $\sigma$, define a matrix $A(\Omega) \in M_n(\mathbb{C})$ by

$$A(\Omega)_{ij} = \begin{cases} 1 & (x_i, x_j) \in \Omega \\ 0 & \text{else.} \end{cases}$$

**Corollary 7.3.3.** *Let $\sigma\colon G \longrightarrow S_X$ be a transitive group action. We retain the above notation. Let $\Omega_1, \ldots, \Omega_r$ be the orbitals of $\sigma$ where $r = \operatorname{rank}(\sigma)$. Then the set $\{A(\Omega_1), \ldots, A(\Omega_r)\}$ is a basis for $C(\sigma)$ and consequently $\dim C(\sigma) = \operatorname{rank}(\sigma)$.*

*Proof.* Proposition 7.2.8 implies that a basis for $\mathbb{C}(X \times X)^G$ is given by the elements $v_1, \ldots, v_r$ where

$$v_k = \sum_{(x_i, x_j) \in \Omega_k} (x_i, x_j).$$

Clearly, $A(\Omega_k) = T^{-1}v_k$. As $T$ restricts to an equivalence of $C(\sigma) = M_n(\mathbb{C})^G$ and $\mathbb{C}(X \times X)^G$ (cf. Exercise 7.7), it follows that $\{A(\Omega_1), \ldots, A(\Omega_r)\}$ is a basis for $C(\sigma)$, as required. □

An important notion in applications is that of a Gelfand pair; the reader is referred to [3, 4.7] and [7, 3.F] where a Fourier transform is defined in this context and applied to probability theory.

**Definition 7.3.4 (Gelfand pair).** Let $G$ be a group and $H$ a subgroup. Let $\sigma\colon G \longrightarrow S_{G/H}$ be the coset action. Then $(G, H)$ is said to be a *Gelfand pair* if the centralizer algebra $C(\sigma)$ is commutative.

*Example 7.3.5.* Let $G$ be a group and let $H = \{1\}$. The coset action of $G$ on $G/H$ is none other than the regular action $\lambda\colon G \longrightarrow S_G$ and so $\widetilde{\lambda}$ is none other than the regular representation $L$. We claim that $C(\lambda) \cong L(G)$. For this argument, we identify the centralizer algebra with the ring $\operatorname{Hom}_G(L, L)$.

Let $T \in C(\lambda)$ and define $f_T\colon G \longrightarrow \mathbb{C}$ by

$$T1 = \sum_{x \in G} f_T(x^{-1})x.$$

We claim that the mapping $T \mapsto f_T$ is an isomorphism $\psi\colon C(\lambda) \longrightarrow L(G)$. First note that, for $g \in G$, one has

$$Tg = TL_g 1 = L_g T1 = L_g \sum_{x \in G} f_T(x^{-1})x.$$

Thus $T$ is determined by $f_T$ and hence $\psi$ is injective. It is also surjective because if $f\colon G \longrightarrow \mathbb{C}$ is any function, then we can define $T \in \operatorname{End}(\mathbb{C}G)$ on the basis by

$$Tg = L_g \sum_{x \in G} f(x^{-1})x.$$

First note that $T$ belongs to the centralizer algebra because if $g, y \in G$, then

$$TL_y g = Tyg = L_{yg} \sum_{x \in G} f(x^{-1})x = L_y Tg.$$

Also, we have

$$T1 = \sum_{x \in G} f(x^{-1})x$$

and so $f_T = f$. Thus $\psi$ is surjective. Finally, we compute, for $T_1, T_2 \in C(\lambda)$,

$$\begin{aligned} T_1T_21 &= T_1 \sum_{x \in G} f_{T_2}(x^{-1})x = \sum_{x \in G} f_{T_2}(x^{-1})T_1L_x1 \\ &= \sum_{x \in G} f_{T_2}(x^{-1})L_x \sum_{y \in G} f_{T_1}(y^{-1})y = \sum_{x,y \in G} f_{T_1}(y^{-1})f_{T_2}(x^{-1})xy. \end{aligned}$$

Setting $g = xy$, $u = x^{-1}$ (and hence $y^{-1} = g^{-1}u^{-1}$) yields

$$T_1T_21 = \sum_{g \in G} \sum_{u \in G} f_{T_1}(g^{-1}u^{-1})f_{T_2}(u)g = \sum_{g \in G} f_{T_1} * f_{T_2}(g^{-1})g.$$

Thus $f_{T_1T_2} = f_{T_1} * f_{T_2}$ and so $\psi$ is a ring homomorphism. We conclude that $\psi$ is an isomorphism.

Consequently, $(G, \{1\})$ is a Gelfand pair if and only if $G$ is abelian because $L(G)$ is commutative if and only if $L(G) = Z(L(G))$. But $\dim Z(L(G)) = |Cl(G)|$ and $\dim L(G) = |G|$, and so $Z(L(G)) = L(G)$ if and only if $G$ is abelian.

It is known that $(G, H)$ is a Gelfand pair if and only if $\widetilde{\sigma}$ is *multiplicity-free*, meaning that each irreducible constituent of $\widetilde{\sigma}$ has multiplicity one [3, Theorem 4.4.2]. We content ourselves here with the special case of so-called symmetric Gelfand pairs.

If $\sigma\colon G \longrightarrow S_X$ is a transitive group action, then to each orbital $\Omega$ of $\sigma$, we can associate its transpose

$$\Omega^T = \{(x_1, x_2) \in X \times X \mid (x_2, x_1) \in \Omega\}.$$

It is easy to see that $\Omega^T$ is indeed an orbital. Let us say that $\Omega$ is *symmetric* if $\Omega = \Omega^T$. For instance, the diagonal orbital $\Delta$ is symmetric. Notice that $A(\Omega^T) = A(\Omega)^T$ and hence $\Omega$ is symmetric if and only if the matrix $A(\Omega)$ is symmetric (and hence self-adjoint, as it has real entries).

**Definition 7.3.6 (Symmetric Gelfand pair).** Let $G$ be a group and $H$ a subgroup with corresponding group action $\sigma\colon G \longrightarrow S_{G/H}$. Then $(G, H)$ is called a *symmetric Gelfand pair* if each orbital of $\sigma$ is symmetric.

Of course, we must show that a symmetric Gelfand pair is indeed a Gelfand pair! First we provide some examples.

*Example 7.3.7.* Let $H \leq G$ and suppose that the action of $G$ on $G/H$ is 2-transitive. Then the orbitals are $\Delta$ and $(G/H \times G/H) \setminus \Delta$. Clearly, each of these is symmetric. Thus $(G, H)$ is a symmetric Gelfand pair.

*Example 7.3.8.* Let $n \geq 2$ and let $[n]^2$ be the set of all two-element subsets of $\{1,\ldots,n\}$. Then $S_n$ acts on $[n]^2$ as follows. Define $\tau\colon S_n \longrightarrow S_{[n]^2}$ by $\tau_\sigma(\{i,j\}) = \{\sigma(i),\sigma(j)\}$. This action is clearly transitive since $S_n$ is 2-transitive on $\{1,\ldots,n\}$. Let $H$ be the stabilizer in $S_n$ of $\{n-1,n\}$. Notice that $H$ is the internal direct product of $S_{n-2}$ and $S_{\{n-1,n\}}$ and so $H \cong S_{n-2} \times S_2$. The action of $S_n$ on $[n]^2$ can be identified with the action of $S_n$ on $S_n/H$.

If $\Omega$ is a non-trivial orbital, then a typical element of $\Omega$ is of the form $(\{i,j\},\{k,\ell\})$ where these two subsets are different. There are essentially two cases. If $i,j,k$, and $\ell$ are all distinct, then $(i\ k)(j\ \ell)$ takes the above element to $(\{k,\ell\},\{i,j\})$ and so $\Omega$ is symmetric. Otherwise, the two subsets have an element in common, say $i = k$. Then $(j\ \ell)$ takes $(\{i,j\},\{i,\ell\})$ to $(\{i,\ell\},\{i,j\})$. Thus $\Omega$ is symmetric in this case, as well. We conclude $(S_n,H)$ is a symmetric Gelfand pair.

The proof that a symmetric Gelfand pair is in fact a Gelfand pair relies on the following simple observation on rings of symmetric matrices.

**Lemma 7.3.9.** *Let $R$ be a subring of $M_n(\mathbb{C})$ consisting of symmetric matrices. Then $R$ is commutative.*

*Proof.* If $A,B \in R$, then $AB = (AB)^T = B^TA^T = BA$ since $A$, $B$, and $AB$ are assumed symmetric. □

And now we turn to the proof that symmetric Gelfand pairs are Gelfand.

**Theorem 7.3.10.** *Let $(G,H)$ be a symmetric Gelfand pair. Then $(G,H)$ is a Gelfand pair.*

*Proof.* As usual, let $\sigma\colon G \longrightarrow S_{G/H}$ be the action map. Denote by $\Omega_1,\ldots,\Omega_r$ the orbitals of $\sigma$. Then because each $\Omega_i$ is symmetric, it follows that each matrix $A(\Omega_i)$ is symmetric for $i = 1,\ldots r$. Since the symmetric matrices form a vector subspace of $M_n(\mathbb{C})$ and $\{A(\Omega_1),\ldots,A(\Omega_r)\}$ is a basis for $C(\sigma)$ by Corollary 7.3.3, it follows that $C(\sigma)$ consists of symmetric matrices. Thus $C(\sigma)$ is commutative by Lemma 7.3.9 and so $(G,H)$ is a Gelfand pair. □

## Exercises

**Exercise 7.1.** Show that if $\sigma\colon G \longrightarrow S_X$ is a group action, then the orbits of $G$ on $X$ form a partition $X$.

**Exercise 7.2.** Let $\sigma\colon G \longrightarrow S_X$ be a transitive group action with $|X| \geq 2$. If $x \in X$, let

$$G_x = \{g \in G \mid \sigma_g(x) = x\}. \tag{7.3}$$

$G_x$ is a subgroup of $G$ called the *stabilizer* of $x$. Prove that the following are equivalent:

1. $G_x$ is transitive on $X \setminus \{x\}$ for *some* $x \in X$;

2. $G_x$ is transitive on $X \setminus \{x\}$ for *all* $x \in X$;
3. $G$ acts 2-transitively on $X$.

**Exercise 7.3.** Compute the character table of $A_4$. (Hints:

1. Let $K = \{Id, (12)(34), (13)(24), (14)(23)\}$. Then $K$ is a normal subgroup of $A_4$ and $A_4/K \cong \mathbb{Z}/3\mathbb{Z}$. Use this to construct 3 degree one representations of $A_4$.
2. Show that $A_4$ acts 2-transitively on $\{1, 2, 3, 4\}$.
3. Conclude that $A_4$ has four conjugacy classes and find them.
4. Produce the character table.)

**Exercise 7.4.** Two group actions $\sigma\colon G \longrightarrow S_X$ and $\tau\colon G \longrightarrow S_Y$ are isomorphic if there is a bijection $\psi\colon X \longrightarrow Y$ such that $\psi\sigma_g = \tau_g\psi$ for all $g \in G$.

1. Show that if $\tau\colon G \longrightarrow S_X$ is a transitive group action, $x \in X$ and $G_x$ is the stabilizer of $x$ (cf. (7.3)), then $\tau$ is isomorphic to the coset action $\sigma\colon G \longrightarrow S_{G/G_x}$.
2. Show that if $\sigma$ and $\tau$ are isomorphic group actions, then the corresponding permutation representations are equivalent.

**Exercise 7.5.** Let $p$ be a prime. Let $G$ be the group of all permutations $\mathbb{Z}/p\mathbb{Z} \longrightarrow \mathbb{Z}/p\mathbb{Z}$ of the form $x \mapsto ax + b$ with $a \in \mathbb{Z}/p\mathbb{Z}^*$ and $b \in \mathbb{Z}/p\mathbb{Z}$. Prove that the action of $G$ on $\mathbb{Z}/p\mathbb{Z}$ is 2-transitive.

**Exercise 7.6.** Let $G$ be a finite group.

1. Suppose that $G$ acts transitively on a finite set $X$ with $|X| \geq 2$. Show that there is an element $g \in G$ with no fixed points on $X$. (Hint: Assume that the statement is false. Use that the identity $e$ has $|X|$ fixed points to contradict Burnside's lemma.)
2. Let $H$ be a proper subgroup of $G$. Prove that

$$G \neq \bigcup_{x \in G} xHx^{-1}.$$

(Hint: Use the previous part.)

**Exercise 7.7.** Let $\varphi\colon G \longrightarrow GL(V)$ and $\rho\colon G \longrightarrow GL(W)$ be representations and suppose that $T\colon V \longrightarrow W$ is an equivalence. Show that $T(V^G) = W^G$ and the restriction $T\colon V^G \longrightarrow W^G$ is an equivalence.

**Exercise 7.8.** Show that if $\Omega$ is an orbital of a transitive group action $\sigma\colon G \longrightarrow S_X$, then the transpose $\Omega^T$ is an orbital of $\sigma$.

**Exercise 7.9.** Suppose that $G$ is a finite group of order $n$ with $s$ conjugacy classes. Suppose that one chooses a pair $(g, h) \in G \times G$ uniformly at random. Prove that the probability $g$ and $h$ commute is $s/n$. (Hint: Apply Burnside's lemma to the action of $G$ on itself by conjugation.)

**Exercise 7.10.** Give a direct combinatorial proof of Burnside's lemma, avoiding character theory.

**Exercise 7.11.** Let $G$ be a group and define $\Lambda\colon G \longrightarrow GL(L(G))$ by putting $\Lambda_g(f)(h) = f(g^{-1}h)$.

1. Verify that $\Lambda$ is a representation.
2. Prove that $\Lambda$ is equivalent to the regular representation $L$.
3. Let $K$ be a subgroup of $G$. Let $L(G/K)$ be the subspace of $L(G)$ consisting of functions $f\colon G \longrightarrow \mathbb{C}$ that are *right $K$-invariant*, that is, $f(gk) = f(g)$ for all $k \in K$. Show that $L(G/K)$ is a $G$-invariant subspace of $L(G)$ and that the restriction of $\Lambda$ to $L(G/K)$ is equivalent to the permutation representation $\mathbb{C}(G/K)$. (Hint: show that $L(G/K)$ has a basis consisting of functions that are constant on left cosets of $K$ and compute the character.)

# Chapter 8
# Induced Representations

If $\psi\colon G \longrightarrow H$ is a group homomorphism, then from any representation $\varphi\colon H \longrightarrow GL(V)$ we can obtain a representation $\rho\colon G \longrightarrow GL(V)$ by composition: set $\rho = \varphi \circ \psi$. If $\psi$ is onto and $\varphi$ is irreducible, one can verify that $\rho$ will also be irreducible. Lemma 6.2.7 shows that every degree one representation of $G$ is obtained in this way by taking $\psi\colon G \longrightarrow G/G'$. As $G/G'$ is abelian, in principle, we know how to compute all its irreducible representations. Now we would like to consider the dual situation: suppose $H$ is a subgroup of $G$; how can we construct a representation of $G$ from a representation of $H$? There is a method to do this, due to Frobenius, via a procedure called induction. This is particularly useful when applied to abelian subgroups because we know how to construct all representations of an abelian group.

## 8.1 Induced Characters and Frobenius Reciprocity

We use the notation $H \leq G$ to indicate that $H$ is a subgroup of $G$. Our goal is to first define the induced character on $G$ associated to a character on $H$. This induced character will be a class function; we will worry later about showing that it is actually the character of a representation. If $f\colon G \longrightarrow \mathbb{C}$ is a function, then we can restrict $f$ to $H$ to obtain a map $\operatorname{Res}^G_H f\colon H \longrightarrow \mathbb{C}$ called the *restriction* of $f$. So $\operatorname{Res}^G_H f(h) = f(h)$ for $h \in H$.

**Proposition 8.1.1.** *Let $H \leq G$. Then $\operatorname{Res}^G_H\colon Z(L(G)) \longrightarrow Z(L(H))$ is a linear map.*

*Proof.* First we need to verify that if $f\colon G \longrightarrow \mathbb{C}$ is a class function, then so is $\operatorname{Res}^G_H f$. Indeed, if $x, h \in H$, then $\operatorname{Res}^G_H f(xhx^{-1}) = f(xhx^{-1}) = f(h) = \operatorname{Res}^G_H f(h)$ since $f$ is a class function. Linearity is immediate:

$$\operatorname{Res}^G_H(c_1f_1 + c_2f_2)(h) = c_1f_1(h) + c_2f_2(h) = c_1 \operatorname{Res}^G_H f_1(h) + c_2 \operatorname{Res}^G_H f_2(h).$$

This completes the proof. □

B. Steinberg, *Representation Theory of Finite Groups: An Introductory Approach*, Universitext, DOI 10.1007/978-1-4614-0776-8_8,
© Springer Science+Business Media, LLC 2012

Our goal now is to construct a linear map $Z(L(H)) \longrightarrow Z(L(G))$ going the other way. First we need a piece of notation. If $H \leq G$ and $f\colon H \longrightarrow \mathbb{C}$ is a function, let us define $\dot{f}\colon G \longrightarrow \mathbb{C}$ by

$$\dot{f}(x) = \begin{cases} f(x) & x \in H \\ 0 & x \notin H. \end{cases}$$

The reader should verify that the assignment $f \mapsto \dot{f}$ is a linear map from $L(H)$ to $L(G)$. Let us now define a map $\operatorname{Ind}_H^G\colon Z(L(H)) \longrightarrow Z(L(G))$, called *induction*, by the formula

$$\operatorname{Ind}_H^G f(g) = \frac{1}{|H|}\sum_{x\in G} \dot{f}(x^{-1}gx).$$

In the case $\chi$ is a character of $H$, one calls $\operatorname{Ind}_H^G \chi$ the *induced character* of $\chi$ on $G$.

**Proposition 8.1.2.** *Let $H \leq G$. Then the map*

$$\operatorname{Ind}_H^G\colon Z(L(H)) \longrightarrow Z(L(G))$$

*is linear.*

*Proof.* First we verify that $\operatorname{Ind}_H^G f$ is really a class function. Let $y, g \in G$. Then one has

$$\operatorname{Ind}_H^G f(y^{-1}gy) = \frac{1}{|H|}\sum_{x\in G} \dot{f}(x^{-1}y^{-1}gyx) = \frac{1}{|H|}\sum_{z\in G} \dot{f}(z^{-1}gz) = \operatorname{Ind}_H^G f(g)$$

where the penultimate equality follows by setting $z = yx$. Next we check linearity. Indeed, we compute

$$\begin{aligned}\operatorname{Ind}_H^G(c_1f_1 + c_2f_2)(g) &= \frac{1}{|H|}\sum_{x\in G} \overbrace{c_1f_2 + c_2f_2}^{\cdot}(x^{-1}gx) \\ &= c_1\frac{1}{|H|}\sum_{x\in G} \dot{f}_1(x^{-1}gx) + c_2\frac{1}{|H|}\sum_{x\in G} \dot{f}_2(x^{-1}gx) \\ &= c_1 \operatorname{Ind}_H^G f_1(g) + c_2 \operatorname{Ind}_H^G f_2(g)\end{aligned}$$

establishing the linearity of the induction map. □

The following theorem, known as Frobenius reciprocity, asserts that the linear maps $\operatorname{Res}_H^G$ and $\operatorname{Ind}_H^G$ are adjoint. What it says is in practice is that if $\chi$ is an irreducible character of $G$ and $\theta$ is an irreducible character of $H$, then the multiplicity of $\chi$ in the induced character $\operatorname{Ind}_H^G \theta$ is exactly the same as the multiplicity of $\theta$ in $\operatorname{Res}_H^G \chi$.

**Theorem 8.1.3 (Frobenius reciprocity).** *Suppose that $H$ is a subgroup of $G$ and let $a$ be a class function on $H$ and $b$ be a class function on $G$. Then the formula*

$$\langle \operatorname{Ind}_H^G a, b\rangle = \langle a, \operatorname{Res}_H^G b\rangle$$

*holds.*

*Proof.* We begin by computing

$$\langle \operatorname{Ind}_H^G a, b\rangle = \frac{1}{|G|}\sum_{g\in G} \operatorname{Ind}_H^G a(g)\overline{b(g)} \tag{8.1}$$

$$= \frac{1}{|G|}\sum_{g\in G}\frac{1}{|H|}\sum_{x\in G}\dot{a}(x^{-1}gx)\overline{b(g)} \tag{8.2}$$

$$= \frac{1}{|G|}\frac{1}{|H|}\sum_{x\in G}\sum_{g\in G}\dot{a}(x^{-1}gx)\overline{b(g)}. \tag{8.3}$$

Now in order for $\dot{a}(x^{-1}gx)$ not to be 0, we need $x^{-1}gx \in H$, that is, we need $g = xhx^{-1}$ with $h \in H$. This allows us to re-index the sum in (8.3) as

$$\begin{aligned}\frac{1}{|G|}\frac{1}{|H|}\sum_{x\in G}\sum_{h\in H} a(h)\overline{b(xhx^{-1})} &= \frac{1}{|G|}\frac{1}{|H|}\sum_{x\in G}\sum_{h\in H} a(h)\overline{b(h)}\\ &= \frac{1}{|G|}\sum_{x\in G}\langle a, \operatorname{Res}_H^G b\rangle\\ &= \langle a, \operatorname{Res}_H^G b\rangle\end{aligned}$$

where the first equality uses that $b$ is a class function on $G$. □

The following formula for induction in terms of coset representatives is often extremely useful, especially for computational purposes.

**Proposition 8.1.4.** *Let $G$ be a group and $H$ a subgroup of $G$. Let $t_1, \dots, t_m$ be a complete set of representatives of the left cosets of $H$ in $G$. Then the formula*

$$\operatorname{Ind}_H^G f(g) = \sum_{i=1}^{m} \dot{f}(t_i^{-1}gt_i)$$

*holds for any class function $f$ on $H$.*

*Proof.* Using that $G$ is the disjoint union $t_1H \cup \cdots \cup t_mH$, we obtain

$$\operatorname{Ind}_H^G f(g) = \frac{1}{|H|}\sum_{x\in G}\dot{f}(x^{-1}gx) = \frac{1}{|H|}\sum_{i=1}^{m}\sum_{h\in H}\dot{f}(h^{-1}t_i^{-1}gt_ih). \tag{8.4}$$

Now if $h \in H$, then $h^{-1}t_i^{-1}gt_ih \in H$ if and only if $t_i^{-1}gt_i \in H$. Since $f$ is a class function on $H$, it follows that $\dot{f}(h^{-1}t_i^{-1}gt_ih) = \dot{f}(t_i^{-1}gt_i)$ and so the right-hand side of (8.4) equals

$$\frac{1}{|H|}\sum_{i=1}^{m}\sum_{h\in H}\dot{f}(t_i^{-1}gt_i) = \frac{1}{|H|}\sum_{h\in H}\sum_{i=1}^{m}\dot{f}(t_i^{-1}gt_i) = \sum_{i=1}^{m}\dot{f}(t_i^{-1}gt_i)$$

completing the proof. □

## 8.2 Induced Representations

If $\varphi\colon G \longrightarrow GL(V)$ is a representation of $G$ and $H \leq G$, then we can restrict $\varphi$ to $H$ to obtain a representation $\operatorname{Res}_H^G \varphi\colon H \longrightarrow GL(V)$. If $h \in H$, then

$$\chi_{\operatorname{Res}_H^G \varphi}(h) = \operatorname{Tr}(\operatorname{Res}_H^G \varphi(h)) = \operatorname{Tr}(\varphi(h)) = \chi_\varphi(h) = \operatorname{Res}_H^G \chi_\varphi(h).$$

It follows that $\chi_{\operatorname{Res}_H^G \varphi} = \operatorname{Res}_H^G \chi_\varphi$. Thus the restriction map sends characters to characters. In this section, we show that induction also sends characters to characters, but the construction in this case is much more complicated. Let us look at some examples to see why this might indeed be the case.

*Example 8.2.1 (Regular representation).* Let $\chi_1$ be the trivial character of the trivial subgroup $\{1\}$ of $G$. Then

$$\operatorname{Ind}_{\{1\}}^G \chi_1(g) = \sum_{x\in G}\dot{\chi}_1(x^{-1}gx),$$

but $x^{-1}gx \in \{1\}$ if and only if $g = 1$. Thus

$$\operatorname{Ind}_{\{1\}}^G \chi_1(g) = \begin{cases} |G| & g = 1 \\ 0 & g \neq 1, \end{cases}$$

i.e., $\operatorname{Ind}_{\{1\}}^G \chi_1$ is the character of the regular representation of $G$.

This example can be generalized.

*Example 8.2.2 (Permutation representations).* Let $H \leq G$ and consider the associated group action $\sigma\colon G \longrightarrow S_{G/H}$ given by $\sigma_g(xH) = gxH$. Notice that $xH \in \operatorname{Fix}(g)$ if and only if $gxH = xH$, that is, $x^{-1}gx \in H$. Now there are $|H|$ elements $x$ giving the coset $xH$ so $|\operatorname{Fix}(g)|$ is $1/|H|$ times the number of $x \in G$ such that $x^{-1}gx \in H$. Let $\chi_1$ be the trivial character of $H$. Then

$$\dot{\chi}_1(x^{-1}gx) = \begin{cases} 1 & x^{-1}gx \in H \\ 0 & x^{-1}gx \notin H \end{cases}$$

and so we deduce

$$\chi_{\widetilde{\sigma}}(g) = |\mathrm{Fix}(g)| = \frac{1}{|H|}\sum_{x\in G}\dot{\chi}_1(x^{-1}gx) = \mathrm{Ind}_H^G\,\chi_1(g)$$

showing that $\mathrm{Ind}_H^G\,\chi_1$ is the character of the permutation representation $\widetilde{\sigma}$.

Fix now a group $G$ and a subgroup $H$. Let $m = [G : H]$ be the index of $H$ in $G$. Choose a complete set of representatives $t_1, \ldots, t_m$ of the left cosets of $H$ in $G$. Without loss of generality we may always take $t_1 = 1$. Suppose $\varphi\colon H \longrightarrow GL_d(\mathbb{C})$ is a representation of $H$. Let us introduce a dot notation in this context by setting

$$\dot{\varphi}_x = \begin{cases} \varphi_x & x \in H \\ 0 & x \notin H \end{cases}$$

where $0$ is the $d \times d$ zero matrix. We now may define a representation $\mathrm{Ind}_H^G\,\varphi\colon G \longrightarrow GL_{md}(\mathbb{C})$, called the *induced representation*, as follows. First, for ease of notation, we write $\varphi^G$ for $\mathrm{Ind}_H^G\,\varphi$. Then, for $g \in G$, we construct $\varphi_g^G$ as an $m \times m$ block matrix with $d \times d$ blocks by setting $[\varphi_g^G]_{ij} = \dot{\varphi}_{t_i^{-1}gt_j}$ for $1 \leq i, j \leq m$. In matrix form we have

$$\varphi_g^G = \begin{bmatrix} \dot{\varphi}_{t_1^{-1}gt_1} & \dot{\varphi}_{t_1^{-1}gt_2} & \cdots & \dot{\varphi}_{t_1^{-1}gt_m} \\ \dot{\varphi}_{t_2^{-1}gt_1} & \dot{\varphi}_{t_2^{-1}gt_2} & \cdots & \vdots \\ \vdots & \vdots & \ddots & \dot{\varphi}_{t_{m-1}^{-1}gt_m} \\ \dot{\varphi}_{t_m^{-1}gt_1} & \cdots & \dot{\varphi}_{t_m^{-1}gt_{m-1}} & \dot{\varphi}_{t_m^{-1}gt_m} \end{bmatrix}.$$

Before proving that $\mathrm{Ind}_H^G\,\varphi$ is indeed a representation, let us look at some examples.

*Example 8.2.3 (Dihedral groups).* Let $G$ be the dihedral group $D_n$ of order $2n$. If $r$ is a rotation by $2\pi/n$ and $s$ is a reflection, then

$$D_n = \{r^m, sr^m \mid 0 \leq m \leq n-1\}.$$

Let $H = \langle r\rangle$; so $H$ is a cyclic subgroup of order $n$ and index 2. For $0 \leq k \leq n-1$, let $\chi_k\colon H \longrightarrow \mathbb{C}^*$ be the representation given by $\chi_k(r^m) = e^{2\pi ikm/n}$. Let us compute the induced representation $\varphi^{(k)} = \mathrm{Ind}_H^G\,\chi_k$. We choose coset representatives $t_1 = 1$ and $t_2 = s$. Then

$$t_1^{-1}r^m t_1 = r^m \qquad t_1^{-1}sr^m t_1 = sr^m$$
$$t_1^{-1}r^m t_2 = r^m s = sr^{-m} \qquad t_1^{-1}sr^m t_2 = sr^m s = r^{-m}$$
$$t_2^{-1}r^m t_1 = sr^m \qquad t_2^{-1}sr^m t_1 = r^m$$
$$t_2^{-1}r^m t_2 = r^{-m} \qquad t_2^{-1}sr^m t_2 = r^m s = sr^{-m}$$

and so we obtain

$$\varphi^{(k)}_{r^m} = \begin{bmatrix} \dot{\chi}_k(r^m) & \dot{\chi}_k(sr^{-m}) \\ \dot{\chi}_k(sr^m) & \dot{\chi}_k(r^{-m}) \end{bmatrix} = \begin{bmatrix} e^{2\pi ikm/n} & 0 \\ 0 & e^{-2\pi ikm/n} \end{bmatrix}$$
$$\varphi^{(k)}_{sr^m} = \begin{bmatrix} \dot{\chi}_k(sr^m) & \dot{\chi}_k(r^{-m}) \\ \dot{\chi}_k(r^m) & \dot{\chi}_k(sr^{-m}) \end{bmatrix} = \begin{bmatrix} 0 & e^{-2\pi ikm/n} \\ e^{2\pi ikm/n} & 0 \end{bmatrix}.$$

In particular, $\mathrm{Ind}_H^G \chi_k(r^m) = 2\cos(2\pi km/n)$ and $\mathrm{Ind}_H^G \chi_k(sr^m) = 0$. It is easy to verify that $\varphi^{(k)}$ is irreducible for $1 \leq k < \frac{n}{2}$ and that this range of values gives inequivalent irreducible representations. Note that $\mathrm{Ind}_H^G \chi_k = \mathrm{Ind}_H^G \chi_{n-k}$, so there is no need to consider $k > \frac{n}{2}$. One can show that the $\varphi^{(k)}$ cover all the equivalence classes of irreducible representations of $D_n$ except for the degree one representations. If $n$ is odd there are two degree one characters while if $n$ is even there are four degree one representations.

*Example 8.2.4 (Quaternions).* Let $Q = \{\pm 1, \pm\hat{\imath}, \pm\hat{\jmath}, \pm\hat{k}\}$ be the group of quaternions. Here $-1$ is central and the rules $\hat{\imath}^2 = \hat{\jmath}^2 = \hat{k}^2 = \hat{\imath}\hat{\jmath}\hat{k} = -1$ are valid. One can verify that $Q' = \{\pm 1\}$ and that $Q/Q' \cong \mathbb{Z}/2\mathbb{Z} \times Z/2\mathbb{Z}$. Thus $Q$ has four degree one representations. Since each representation has degree dividing 8 and the sum of the squares of the degrees is 8, there can be only one remaining irreducible representation and it must have degree 2. Let us construct it as an induced representation. Let $H = \langle \hat{\imath} \rangle$. Then $|H| = 4$ and so $[Q : H] = 2$. Consider the representation $\varphi\colon H \longrightarrow \mathbb{C}^*$ given by $\varphi(\hat{\imath}^k) = i^k$. Let $t_1 = 1$ and $t_2 = \hat{\jmath}$. Then one can compute

$$\varphi^Q_{\pm 1} = \pm \begin{bmatrix} 1 & 0 \\ 0 & 1 \end{bmatrix}, \qquad \varphi^Q_{\pm\hat{\imath}} = \pm \begin{bmatrix} i & 0 \\ 0 & -i \end{bmatrix},$$
$$\varphi^Q_{\pm\hat{\jmath}} = \pm \begin{bmatrix} 0 & -1 \\ 1 & 0 \end{bmatrix}, \qquad \varphi^Q_{\pm\hat{k}} = \pm \begin{bmatrix} 0 & -i \\ -i & 0 \end{bmatrix}.$$

It is easy to see that $\varphi^Q$ is irreducible since $\varphi^Q_{\hat{\imath}}$ and $\varphi^Q_{\hat{\jmath}}$ have no common eigenvector. The character table of $Q$ appears in Table 8.1.

We are now ready to prove that $\mathrm{Ind}_H^G \varphi$ is a representation with character $\mathrm{Ind}_H^G \chi_\varphi$.

**Table 8.1** Character table of the quaternions

| | $1$ | $-1$ | $\hat{\imath}$ | $\hat{\jmath}$ | $\hat{k}$ |
|---|---|---|---|---|---|
| $\chi_1$ | 1 | 1 | 1 | 1 | 1 |
| $\chi_2$ | 1 | 1 | 1 | −1 | −1 |
| $\chi_3$ | 1 | 1 | −1 | 1 | −1 |
| $\chi_4$ | 1 | 1 | −1 | −1 | 1 |
| $\chi_5$ | 2 | −2 | 0 | 0 | 0 |

**Theorem 8.2.5.** *Let $H$ be a subgroup of $G$ of index $m$ and suppose that $\varphi\colon H \longrightarrow GL_d(\mathbb{C})$ is a representation of $H$. Then $\mathrm{Ind}_H^G\,\varphi\colon G \longrightarrow GL_{md}(\mathbb{C})$ is a representation and $\chi_{\mathrm{Ind}_H^G\varphi} = \mathrm{Ind}_H^G\,\chi_\varphi$. In particular, $\mathrm{Ind}_H^G$ maps characters to characters.*

*Proof.* Let $t_1, \ldots, t_m$ be a complete set of representatives for the cosets of $H$ in $G$. Set $\varphi^G = \mathrm{Ind}_H^G\,\varphi$. We begin by showing that $\varphi^G$ is a representation. Let $x, y \in G$. Then we have

$$[\varphi_x^G\varphi_y^G]_{ij} = \sum_{k=1}^{m}[\varphi_x^G]_{ik}[\varphi_y^G]_{kj} = \sum_{k=1}^{m}\dot{\varphi}_{t_i^{-1}xt_k}\dot{\varphi}_{t_k^{-1}yt_j}. \tag{8.5}$$

The only way $\dot{\varphi}_{t_k^{-1}yt_j} \neq 0$ is if $t_k^{-1}yt_j \in H$, or equivalently $t_kH = yt_jH$. So if $t_\ell$ is the representative of the coset $yt_jH$, then the right-hand side of (8.5) becomes $\dot{\varphi}_{t_i^{-1}xt_\ell}\varphi_{t_\ell^{-1}yt_j}$. This in turn is non-zero if and only if $t_i^{-1}xt_\ell \in H$, that is, $t_iH = xt_\ell H = xyt_jH$ or equivalently $t_i^{-1}xyt_j \in H$. If this is the case, then the right-hand side of (8.5) equals

$$\varphi_{t_i^{-1}xt_\ell}\varphi_{t_\ell^{-1}yt_j} = \varphi_{t_i^{-1}xyt_j}$$

and hence $[\varphi_x^G\varphi_y^G]_{ij} = \dot{\varphi}_{t_i^{-1}xyt_j} = [\varphi_{xy}^G]_{ij}$, establishing that $\varphi^G$ preserves multiplication. Next observe that $[\varphi_1^G]_{ij} = \dot{\varphi}_{t_i^{-1}t_j}$, but $t_i^{-1}t_j \in H$ implies $t_iH = t_jH$, which in turn implies $t_i = t_j$. Thus

$$[\varphi_1^G]_{ij} = \begin{cases}\varphi_1 = I & i = j\\ 0 & i \neq j\end{cases}$$

and so $\varphi_1^G = I$. Therefore, if $g \in G$ then $\varphi_g^G\varphi_{g^{-1}}^G = \varphi_{gg^{-1}}^G = \varphi_1^G = I$ establishing that $(\varphi_g^G)^{-1} = \varphi_{g^{-1}}^G$ and therefore $\varphi^G$ is a representation. Let us compute its character.

Applying Proposition 8.1.4 we obtain

$$\chi_{\varphi^G}(g) = \mathrm{Tr}(\varphi_g^G) = \sum_{i=1}^{m}\mathrm{Tr}(\dot{\varphi}_{t_i^{-1}gt_i}) = \sum_{i=1}^{m}\dot{\chi}_\varphi(t_i^{-1}gt_i) = \mathrm{Ind}_H^G\,\chi_\varphi$$

as required. □

## 8.3 Mackey's Irreducibility Criterion

There is no guarantee that if $\chi$ is an irreducible character of $H$, then $\mathrm{Ind}_H^G \chi$ will be an irreducible character of $G$. For instance, $L = \mathrm{Ind}_{\{1\}}^G \chi_1$ is not irreducible. On the other hand, sometimes induced characters are irreducible, as was the case for the representations of the dihedral groups and the quaternions considered in the examples. There is a criterion, due to Mackey, describing when an induced character is irreducible. This is the subject of this section. By Frobenius reciprocity,

$$\langle \mathrm{Ind}_H^G \chi_\varphi, \mathrm{Ind}_H^G \chi_\varphi \rangle = \langle \chi_\varphi, \mathrm{Res}_H^G \mathrm{Ind}_H^G \chi_\varphi \rangle$$

and so our problem amounts to understanding $\mathrm{Res}_H^G \mathrm{Ind}_H^G \chi_\varphi$.

**Definition 8.3.1 (Disjoint representations).** Two representations $\varphi$ and $\rho$ of $G$ are said to be *disjoint* if they have no common irreducible constituents.

**Proposition 8.3.2.** *Representations $\varphi$ and $\rho$ of $G$ are disjoint if and only if $\chi_\varphi$ and $\chi_\rho$ are orthogonal.*

*Proof.* Let $\varphi^{(1)}, \ldots, \varphi^{(s)}$ be a complete set of representatives of the equivalence classes of irreducible representations of $G$. Then

$$\varphi \sim m_1\varphi^{(1)} + \cdots + m_s\varphi^{(s)}$$
$$\rho \sim n_1\varphi^{(1)} + \cdots + n_s\varphi^{(s)}$$

for certain non-negative integers $m_i, n_i$. From the orthonormality of irreducible characters, we obtain

$$\langle \chi_\varphi, \chi_\rho \rangle = m_1n_1 + \cdots + m_sn_s. \tag{8.6}$$

Clearly the right-hand side of (8.6) is 0 if and only if $m_in_i = 0$ all $i$, if and only if $\varphi$ and $\rho$ are disjoint. □

The problem of understanding $\mathrm{Res}_H^G \mathrm{Ind}_H^G f$ turns out not to be much more difficult than analyzing $\mathrm{Res}_H^G \mathrm{Ind}_K^G f$ where $H, K$ are two subgroups of $G$. To perform this analysis we need the notion of a double coset.

**Definition 8.3.3 (Double coset).** Let $H, K$ be subgroups of a group $G$. Then define a group action $\sigma\colon H \times K \longrightarrow G$ by $\sigma_{(h,k)}(g) = hgk^{-1}$. The orbit of $g$ under $H \times K$ is then the set

$$HgK = \{hgk \mid h \in H, k \in K\}$$

and is called a *double coset* of $g$. We write $H\backslash G/K$ for the set of double cosets of $H$ and $K$ in $G$.

Notice that the double cosets are disjoint and have union $G$. Also, if $H$ is a normal subgroup of $G$, then $H\backslash G/H = G/H$.

*Example 8.3.4.* Let $G = GL_2(\mathbb{C})$ and $B$ be the group of invertible $2 \times 2$ upper triangular matrices over $\mathbb{C}$. Then $B\backslash G/B = \left\{B, B\begin{bmatrix}0&1\\1&0\end{bmatrix}B\right\}$.

The following theorem of Mackey explains how induction and restriction of class functions from different subgroups interact.

**Theorem 8.3.5 (Mackey).** *Let $H, K \leq G$ and let $S$ be a complete set of double coset representatives for $H\backslash G/K$. Then, for $f \in Z(L(K))$,*

$$\operatorname{Res}^G_H \operatorname{Ind}^G_K f = \sum_{s\in S} \operatorname{Ind}^H_{H\cap sKs^{-1}} \operatorname{Res}^{sKs^{-1}}_{H\cap sKs^{-1}} f^s$$

*where $f^s \in Z(L(sKs^{-1}))$ is given by $f^s(x) = f(s^{-1}xs)$.*

*Proof.* For this proof it is important to construct the "correct" set $T$ of left coset representatives for $K$ in $G$. Choose, for each $s \in S$, a complete set $V_s$ of representatives of the left cosets of $H \cap sKs^{-1}$ in $H$. Then

$$H = \bigcup_{v\in V_S} v(H \cap sKs^{-1})$$

and the union is disjoint. Now

$$HsK = HsKs^{-1}s = \bigcup_{v\in V_s} v(H \cap sKs^{-1})sKs^{-1}s = \bigcup_{v\in V_s} vsK$$

and moreover this union is disjoint. Indeed, if $vsK = v'sK$ with $v, v' \in V_s$, then $s^{-1}v^{-1}v's \in K$ and so $v^{-1}v' \in sKs^{-1}$. But also $v, v' \in H$ so $v^{-1}v' \in H \cap sKs^{-1}$ and hence $v(H\cap sKs^{-1}) = v'(H\cap sKs^{-1})$. Thus $v = v'$ by definition of $V_s$.

Let $T_s = \{vs \mid v \in V_s\}$ and let $T = \bigcup_{s\in S} T_s$. This latter union is disjoint since if $vs = v's'$ for $v \in V_s$ and $v' \in V_{s'}$, then $HsK = Hs'K$ and so $s = s'$, as $S$ is a complete set of double coset representatives. Consequently, $v = v'$. Putting this all together we have

$$G = \bigcup_{s\in S} HsK = \bigcup_{s\in S}\bigcup_{v\in V_s} vsK = \bigcup_{s\in S}\bigcup_{t\in T_s} tK = \bigcup_{t\in T} tK$$

and all these unions are disjoint. Therefore, $T$ is a complete set of representatives for the left cosets of $K$ in $G$.

Using twice Proposition 8.1.4 we compute, for $h \in H$,

$$\begin{aligned}\operatorname{Ind}^G_K f(h) &= \sum_{t\in T} \dot{f}(t^{-1}ht)\\ &= \sum_{s\in S}\sum_{t\in T_s} \dot{f}(t^{-1}ht)\end{aligned}$$

$$= \sum_{s\in S}\sum_{v\in V_s} \dot{f}(s^{-1}v^{-1}hvs)$$

$$= \sum_{s\in S} \sum_{\substack{v\in V_s,\\ v^{-1}hv\in sKs^{-1}}} f^s(v^{-1}hv)$$

$$= \sum_{s\in S} \sum_{\substack{v\in V_s,\\ v^{-1}hv\in H\cap sKs^{-1}}} \mathrm{Res}^{sKs^{-1}}_{H\cap sKs^{-1}} f^s(v^{-1}hv)$$

$$= \sum_{s\in S} \mathrm{Ind}^{H}_{H\cap sKs^{-1}} \mathrm{Res}^{sKs^{-1}}_{H\cap sKs^{-1}} f^s.$$

This completes the proof. □

From Theorem 8.3.5, we deduce Mackey's irreducibility criterion.

**Theorem 8.3.6 (Mackey's irreducibility criterion).** *Let $H$ be a subgroup of $G$ and let $\varphi\colon H \longrightarrow GL_d(\mathbb{C})$ be a representation. Then $\mathrm{Ind}_H^G \varphi$ is irreducible if and only if:*

1. *$\varphi$ is irreducible;*
2. *the representations $\mathrm{Res}^{H}_{H\cap sHs^{-1}} \varphi$ and $\mathrm{Res}^{sHs^{-1}}_{H\cap sHs^{-1}} \varphi^s$ are disjoint for all $s \notin H$, where $\varphi^s(x) = \varphi(s^{-1}xs)$ for $x \in sHs^{-1}$.*

*Proof.* Let $\chi$ be the character of $\varphi$. Let $S$ be a complete set of double coset representatives of $H\backslash G/H$. Assume without loss of generality that $1 \in S$. Then, for $s = 1$, notice that $H \cap sHs^{-1} = H$, $\varphi^s = \varphi$. Let $S^\sharp = S \setminus \{1\}$. Theorem 8.3.5 then yields

$$\mathrm{Res}^G_H \mathrm{Ind}^G_H \chi = \chi + \sum_{s\in S^\sharp} \mathrm{Ind}^{H}_{H\cap sHs^{-1}} \mathrm{Res}^{sHs^{-1}}_{H\cap sHs^{-1}} \chi^s.$$

Applying Frobenius reciprocity twice, we obtain

$$\begin{aligned}\langle \mathrm{Ind}^G_H \chi, \mathrm{Ind}^G_H \chi\rangle &= \langle \mathrm{Res}^G_H \mathrm{Ind}^G_H \chi, \chi\rangle \\ &= \langle \chi,\chi\rangle + \sum_{s\in S^\sharp} \langle \mathrm{Ind}^{H}_{H\cap sHs^{-1}} \mathrm{Res}^{sHs^{-1}}_{H\cap sHs^{-1}} \chi^s, \chi\rangle \\ &= \langle \chi,\chi\rangle + \sum_{s\in S^\sharp} \langle \mathrm{Res}^{sHs^{-1}}_{H\cap sHs^{-1}} \chi^s, \mathrm{Res}^{H}_{H\cap sHs^{-1}} \chi\rangle\end{aligned}$$

where the first equality uses that the inner product in question is real-valued. Since $\langle \chi,\chi\rangle \geq 1$ and all terms in the sum are non-negative, we see that the inner product $\langle \mathrm{Ind}^G_H \chi, \mathrm{Ind}^G_H \chi\rangle$ is 1 if and only if $\langle \chi,\chi\rangle = 1$ and

$$\langle \mathrm{Res}^{sHs^{-1}}_{H\cap sHs^{-1}} \chi^s, \mathrm{Res}^{H}_{H\cap sHs^{-1}} \chi\rangle = 0$$

all $s \in S^{\sharp}$. Thus $\mathrm{Ind}_H^G \varphi$ is irreducible if and only if $\varphi$ is irreducible and the representations $\mathrm{Res}^{sHs^{-1}}_{H\cap sHs^{-1}} \varphi^s$ and $\mathrm{Res}^{H}_{H\cap sHs^{-1}} \varphi$ are disjoint for all $s \in S^{\sharp}$ (using Proposition 8.3.2). Now any $s \notin H$ can be an element of $S^{\sharp}$ for an appropriately chosen set $S$ of double coset representatives, from which the theorem follows. □

*Remark 8.3.7.* The proof shows that one needs to check only that the second condition holds for all $s \notin H$ from a given set of double coset representatives, which is often easier to deal with in practice.

Mackey's criterion is most readily applicable in the case that $H$ is a normal subgroup. In this setting one has the simplifications $H\backslash G/H = G/H$ and $H \cap sHs^{-1} = H$. So Mackey's criterion in this case boils down to checking that $\varphi\colon H \longrightarrow GL_d(\mathbb{C})$ is irreducible and that $\varphi^s\colon H \longrightarrow GL_d(\mathbb{C})$ does not have $\varphi$ as an irreducible constituent for $s \notin H$. Actually, one can easily verify that $\varphi^s$ is irreducible if and only if $\varphi$ is irreducible (see Exercise 8.6), so basically one just has to check that $\varphi$ and $\varphi^s$ are inequivalent irreducible representations when $s \notin H$. In fact, one just needs to check this as $s$ ranges over a complete set of coset representatives for $G/H$.

*Example 8.3.8.* Let $p$ be a prime and let

$$G = \left\{ \begin{bmatrix} [a] & [b] \\ [0] & [1] \end{bmatrix} \mid [a] \in \mathbb{Z}/p\mathbb{Z}^*, [b] \in \mathbb{Z}/p\mathbb{Z} \right\},$$

$$H = \left\{ \begin{bmatrix} [1] & [b] \\ [0] & [1] \end{bmatrix} \mid [b] \in \mathbb{Z}/p\mathbb{Z} \right\}.$$

Then $H \cong \mathbb{Z}/p\mathbb{Z}$, $H \lhd G$, and $G/H \cong \mathbb{Z}/p\mathbb{Z}^*$ (consider the projection to the upper left corner). A complete set of coset representatives for $G/H$ is

$$S = \left\{ \begin{bmatrix} [a] & [0] \\ [0] & [1] \end{bmatrix} \mid [a] \in \mathbb{Z}/p\mathbb{Z}^* \right\}.$$

Let $\varphi\colon H \longrightarrow \mathbb{C}^*$ be given by

$$\varphi\left( \begin{bmatrix} [1] & [b] \\ [0] & [1] \end{bmatrix} \right) = e^{2\pi i b/p}.$$

Then if $s = \begin{bmatrix} [a]^{-1} & [0] \\ [0] & [1] \end{bmatrix}$ with $[a] \neq 1$, we have

$$\varphi^s\left( \begin{bmatrix} [1] & [b] \\ [0] & [1] \end{bmatrix} \right) = \varphi\left( \begin{bmatrix} [1] & [a][b] \\ [0] & [1] \end{bmatrix} \right) = e^{2\pi i ab/p}$$

and so $\varphi, \varphi^s$ are inequivalent irreducible representations of $H$. Mackey's criterion now implies that $\mathrm{Ind}_H^G \varphi$ is an irreducible representation of $G$ of degree $[G : H] = p-1$. Notice that

$$p-1+(p-1)^2 = (p-1)[1+p-1] = (p-1)p = |G|.$$

Since one can lift the $p-1$ degree one representations of $G/H \cong \mathbb{Z}/p\mathbb{Z}^*$ to $G$, the above computation implies that $\mathrm{Ind}_H^G \varphi$ and the $p-1$ degree one representations are all the irreducible representations of $G$.

## Exercises

**Exercise 8.1.** Prove that if $G = GL_2(\mathbb{C})$ and $B = \left\{ \begin{bmatrix} a & b \\ 0 & c \end{bmatrix} \mid ac \neq 0 \right\}$, then $B \backslash G/B = \left\{ B, B \begin{bmatrix} 0 & 1 \\ 1 & 0 \end{bmatrix} B \right\}$. Prove that $G/B$ is infinite.

**Exercise 8.2.** Let $H, K$ be subgroups of $G$. Let $\sigma \colon H \longrightarrow S_{G/K}$ be given by $\sigma_h(gK) = hgK$. Show that the set of double cosets $H \backslash G/K$ is the set of orbits of $H$ on $G/K$.

**Exercise 8.3.** Let $G$ be a group with a cyclic normal subgroup $H = \langle a \rangle$ of order $k$. Suppose that $N_G(a) = H$, that is, $sa = as$ implies $s \in H$. Show that if $\chi \colon H \longrightarrow \mathbb{C}^*$ is the character given by $\chi(a^m) = e^{2\pi i m/k}$, then $\mathrm{Ind}_H^G \chi$ is an irreducible character of $G$.

**Exercise 8.4.**

1. Construct the character table for the dihedral group $D_4$ of order 8. Suppose that $s$ is the reflection over the $x$-axis and $r$ is rotation by $\pi/2$. (Hint: Observe that $Z = \{1, r^2\}$ is in the center of $D_4$ (actually, it is the center) and $D_4/Z \cong \mathbb{Z}/2\mathbb{Z} \times \mathbb{Z}/2\mathbb{Z}$. Use this to get the degree one characters. Get the degree two character as an induced character from a representation of the subgroup $\langle r \rangle \cong \mathbb{Z}/4\mathbb{Z}$.)
2. Is the action of $D_4$ on the vertices of the square 2-transitive?

**Exercise 8.5.** Compute the character table for the group in Example 8.3.8 when $p = 5$.

**Exercise 8.6.** Let $N$ be a normal subgroup of a group $G$ and suppose that $\varphi \colon N \longrightarrow GL_d(\mathbb{C})$ is a representation. For $s \in G$, define $\varphi^s \colon N \longrightarrow GL_d(\mathbb{C})$ by $\varphi^s(n) = \varphi(s^{-1}ns)$. Prove that $\varphi$ is irreducible if and only if $\varphi^s$ is irreducible.

**Exercise 8.7.** Show that if $G$ is a non-abelian group and $\varphi \colon Z(G) \longrightarrow GL_d(\mathbb{C})$ is an irreducible representation of the center of $G$, then $\mathrm{Ind}_{Z(G)}^G \varphi$ is not irreducible.

**Exercise 8.8.** A representation $\varphi \colon G \longrightarrow GL(V)$ is called *faithful* if it is one-to-one.

1. Let $H$ be a subgroup of $G$ and suppose $\varphi\colon H \longrightarrow GL_d(\mathbb{C})$ is a faithful representation. Show that $\varphi^G = \mathrm{Ind}_H^G\,\varphi$ is a faithful representation of $G$.
2. Show that every representation of a simple group which is not a direct sum of copies of the trivial representation is faithful.

**Exercise 8.9.** Let $G$ be a group and let $H$ be a subgroup. Let $\sigma\colon G \longrightarrow S_{G/H}$ be the group action given by $\sigma_g(xH) = gxH$.

1. Show that $\sigma$ is transitive.
2. Show that $H$ is the stabilizer of the coset $H$.
3. Recall that if $1_H$ is the trivial character of $H$, then $\mathrm{Ind}_H^G\,1_H$ is the character $\chi_{\widetilde{\sigma}}$ of the permutation representation $\widetilde{\sigma}\colon G \longrightarrow GL(\mathbb{C}(G/H))$. Use Frobenius reciprocity to show that the rank of $\sigma$ is the number of orbits of $H$ on $G/H$.
4. Conclude that $G$ is 2-transitive on $G/H$ if and only if $H$ is transitive on the set of cosets not equal to $H$ in $G/H$.
5. Show that the rank of $\sigma$ is also the number of double cosets in $H\backslash G/H$ either directly or by using Mackey's Theorem.

**Exercise 8.10.** Use Frobenius reciprocity to give another proof that if $\rho$ is an irreducible representation of $G$, then the multiplicity of $\rho$ as a constituent of the regular representation is the degree of $\rho$.

**Exercise 8.11.** Let $G$ be a group and $H$ a subgroup. Given a representation $\rho\colon H \longrightarrow GL(V)$, let $\mathrm{Hom}_H(G,V)$ be the vector space of all functions $f\colon G \longrightarrow V$ such that $f(gh) = \rho(h)^{-1}f(g)$, for all $g \in G$ and $h \in H$, equipped with pointwise operations. Define a representation

$$\varphi\colon G \longrightarrow GL(\mathrm{Hom}_H(G,V))$$

by $\varphi_g(f)(g_0) = f(g^{-1}g_0)$. Prove that $\varphi$ is a representation of $G$ equivalent to $\mathrm{Ind}_H^G\,\rho$. (Hint: find a basis for $\mathrm{Hom}_H(G,V)$ and compute the character.)

**Exercise 8.12.** Let $G$ be a group and let $H$ be a subgroup with corresponding group action $\sigma\colon G \longrightarrow S_{G/H}$. Define

$$L(H\backslash G/H) = \{f \in L(G) \mid f(h_1gh_2) = f(g), \forall h_1, h_2 \in H, g \in G\}$$

to be the set of $H$-bi-invariant functions.

1. Show that $L(H\backslash G/H)$ is a unital subring of $L(G)$.
2. Prove that $L(H\backslash G/H)$ is isomorphic to the centralizer algebra $C(\sigma)$. (Hint: Use Exercise 7.11 to identify $\mathbb{C}(G/H)$ with $L(G/H)$ and let $L(H\backslash G/H)$ act on $L(G/H)$ by convolution.)
3. Show that $(G,H)$ is a symmetric Gelfand pair if and only if $g^{-1} \in HgH$ for all $g \in G$.

# Chapter 9
# Another Theorem of Burnside

In this chapter we give another application of representation theory to finite group theory, again due to Burnside. The result is based on a study of real characters and conjugacy classes.

## 9.1 Conjugate Representations

Recall that if $A = (a_{ij})$ is a matrix, then $\overline{A}$ is the matrix $(\overline{a_{ij}})$. One easily verifies that $\overline{AB} = \overline{A} \cdot \overline{B}$ and that if $A$ is invertible, then so is $\overline{A}$ and moreover $\overline{A}^{-1} = \overline{A^{-1}}$. Hence if $\varphi\colon G \longrightarrow GL_d(\mathbb{C})$ is a representation of $G$, then we can define the *conjugate representation* $\overline{\varphi}$ by $\overline{\varphi}_g = \overline{\varphi_g}$. If $f\colon G \longrightarrow \mathbb{C}$ is a function, then define $\overline{f}$ by $\overline{f}(g) = \overline{f(g)}$.

**Proposition 9.1.1.** *Let $\varphi\colon G \longrightarrow GL_d(\mathbb{C})$ be a representation. Then we have $\chi_{\overline{\varphi}} = \overline{\chi_\varphi}$.*

*Proof.* First note that if $A \in M_d(\mathbb{C})$, then

$$\mathrm{Tr}(\overline{A}) = \overline{a_{11}} + \cdots + \overline{a_{dd}} = \overline{a_{11} + \cdots + a_{dd}} = \overline{\mathrm{Tr}(A)}.$$

Thus $\chi_{\overline{\varphi}}(g) = \mathrm{Tr}(\overline{\varphi_g}) = \overline{\mathrm{Tr}(\varphi_g)} = \overline{\chi_\varphi(g)}$, as required. □

As a consequence, we observe that the conjugate of an irreducible representation is again irreducible.

**Corollary 9.1.2.** *Let $\varphi\colon G \longrightarrow GL_d(\mathbb{C})$ be irreducible. Then $\overline{\varphi}$ is irreducible.*

*Proof.* Let $\chi = \chi_\varphi$. We compute

$$\langle \overline{\chi}, \overline{\chi} \rangle = \frac{1}{|G|} \sum_{g \in G} \overline{\overline{\chi(g)}}\,\overline{\chi(g)} = \frac{1}{|G|} \sum_{g \in G} \chi(g)\overline{\chi(g)} = \langle \chi, \chi \rangle = 1$$

and so $\overline{\varphi}$ is irreducible by Proposition 9.1.1 and Corollary 4.3.15. □

B. Steinberg, *Representation Theory of Finite Groups: An Introductory Approach*, Universitext, DOI 10.1007/978-1-4614-0776-8_9,
© Springer Science+Business Media, LLC 2012

Quite often one can use the above corollary to produce new irreducible characters for a group. However, the case when $\overline{\chi} = \chi$ is also of importance.

**Definition 9.1.3 (Real character).** A character $\chi$ of $G$ is called *real*[1] if $\chi = \overline{\chi}$, that is, $\chi(g) \in \mathbb{R}$ for all $g \in G$.

*Example 9.1.4.* The trivial character of a group is always real. The groups $S_3$ and $S_4$ have only real characters. On the other hand, if $n$ is odd then $\mathbb{Z}/n\mathbb{Z}$ has no non-trivial real characters.

Since the number of irreducible characters equals the number of conjugacy classes, there should be a corresponding notion of a "real" conjugacy class. First we make two simple observations.

**Proposition 9.1.5.** *Let $\chi$ be a character of a group $G$. Then $\chi(g^{-1}) = \overline{\chi(g)}$.*

*Proof.* Without loss of generality, we may assume that $\chi$ is the character of a unitary representation $\varphi\colon G \longrightarrow U_n(\mathbb{C})$. Then

$$\chi(g^{-1}) = \mathrm{Tr}(\varphi_{g^{-1}}) = \mathrm{Tr}(\overline{\varphi_g}^T) = \mathrm{Tr}(\overline{\varphi_g}) = \overline{\mathrm{Tr}(\varphi_g)} = \overline{\chi(g)}$$

as required. □

**Proposition 9.1.6.** *Let $g$ and $h$ be conjugate. Then $g^{-1}$ and $h^{-1}$ are conjugate.*

*Proof.* Suppose $g = xhx^{-1}$. Then $g^{-1} = xh^{-1}x^{-1}$. □

So if $C$ is a conjugacy class of $G$, then $C^{-1} = \{g^{-1} \mid g \in C\}$ is also a conjugacy class of $G$ and moreover if $\chi$ is any character then $\chi(C^{-1}) = \overline{\chi(C)}$.

**Definition 9.1.7 (Real conjugacy class).** A conjugacy class $C$ of $G$ is said to be *real* if $C = C^{-1}$.

The following proposition motivates the name.

**Proposition 9.1.8.** *Let $C$ be a real conjugacy class and $\chi$ a character of $G$. Then $\chi(C) = \overline{\chi(C)}$, that is, $\chi(C) \in \mathbb{R}$.*

*Proof.* If $C$ is real then $\chi(C) = \chi(C^{-1}) = \overline{\chi(C)}$. □

An important result of Burnside is that the number of real irreducible characters is equal to the number of real conjugacy classes. The elegant proof we provide is due to R. Brauer and is based on the invertibility of the character table. First we prove a lemma.

**Lemma 9.1.9.** *Let $\varphi\colon S_n \longrightarrow GL_n(\mathbb{C})$ be the standard representation of $S_n$ and let $A \in M_n(\mathbb{C})$ be a matrix. Then, for $\sigma \in S_n$, the matrix $\varphi_\sigma A$ is obtained from*

---

[1]Some authors divide what we call real characters into two subclasses: real characters and quaternionic characters.

*A by permuting the rows of A according to $\sigma$ and $A\varphi_\sigma$ is obtained from A by permuting the columns of A according to $\sigma^{-1}$.*

*Proof.* We compute $(\varphi_\sigma A)_{\sigma(i)j} = \sum_{k=1}^{n} \varphi(\sigma)_{\sigma(i)k} A_{kj} = A_{ij}$ since

$$\varphi(\sigma)_{\sigma(i)k} = \begin{cases} 1 & k = i \\ 0 & \text{else.} \end{cases}$$

Thus $\varphi_\sigma A$ is obtained from $A$ by placing row $i$ of $A$ into row $\sigma(i)$. Since the representation $\varphi$ is unitary and real-valued, $A\varphi_\sigma = (\varphi_\sigma^T A^T)^T = (\varphi_{\sigma^{-1}} A^T)^T$ the second statement follows from the first. □

**Theorem 9.1.10 (Burnside).** *Let G be a finite group. The number of real irreducible characters of G equals the number of real conjugacy classes of G.*

*Proof (Brauer).* Let $s$ be the number of conjugacy classes of $G$. Our standing notation will be that $\chi_1, \ldots, \chi_s$ are the irreducible characters of $G$ and $C_1, \ldots, C_s$ are the conjugacy classes. Define $\alpha, \beta \in S_s$ by $\overline{\chi_i} = \chi_{\alpha(i)}$ and $C_i^{-1} = C_{\beta(i)}$. Notice that $\chi_i$ is a real character if and only if $\alpha(i) = i$ and similarly $C_i$ is a real conjugacy class if and only if $\beta(i) = i$. Therefore, $|\mathrm{Fix}(\alpha)|$ is the number of real irreducible characters and $|\mathrm{Fix}(\beta)|$ is the number of real conjugacy classes. Notice that $\alpha = \alpha^{-1}$ since $\alpha$ swaps the indices of $\chi_i$ and $\overline{\chi_i}$, and similarly $\beta = \beta^{-1}$.

Let $\varphi\colon S_s \longrightarrow GL_s(\mathbb{C})$ be the standard representation of $S_s$. Then we have $\chi_\varphi(\alpha) = |\mathrm{Fix}(\alpha)|$ and $\chi_\varphi(\beta) = |\mathrm{Fix}(\beta)|$ so it suffices to prove $\mathrm{Tr}(\varphi_\alpha) = \mathrm{Tr}(\varphi_\beta)$. Let $\mathsf{X}$ be the character table of $G$. Then, by Lemma 9.1.9, $\varphi_\alpha \mathsf{X}$ is obtained from $\mathsf{X}$ by swapping the rows of $\mathsf{X}$ corresponding to $\chi_i$ and $\overline{\chi_i}$ for each $i$. But this means that $\varphi_\alpha \mathsf{X} = \overline{\mathsf{X}}$. Similarly, $\mathsf{X}\varphi_\beta$ is obtained from $\mathsf{X}$ by swapping the columns of $\mathsf{X}$ corresponding to $C_i$ and $C_i^{-1}$ for each $i$. Since $\chi(C^{-1}) = \overline{\chi(C)}$ for each conjugacy class $C$, this swapping again results in $\overline{\mathsf{X}}$. In other words,

$$\varphi_\alpha \mathsf{X} = \overline{\mathsf{X}} = \mathsf{X}\varphi_\beta.$$

But by the second orthogonality relations (Theorem 4.4.12) the columns of $\mathsf{X}$ form an orthogonal set of non-zero vectors and hence are linearly independent. Thus $\mathsf{X}$ is invertible and so $\varphi_\alpha = \mathsf{X}\varphi_\beta \mathsf{X}^{-1}$. We conclude $\mathrm{Tr}(\varphi_\alpha) = \mathrm{Tr}(\varphi_\beta)$, as was required. □

As a consequence, we see that groups of odd order do not have non-trivial real irreducible characters.

**Proposition 9.1.11.** *Let G be a group. Then $|G|$ is odd if and only if G does not have any non-trivial real irreducible characters.*

*Proof.* By Theorem 9.1.10, it suffices to show that $\{1\}$ is the only real conjugacy class of $G$ if and only if $|G|$ is odd. Suppose first $G$ has even order. Then there is an element $g \in G$ of order 2. Since $g = g^{-1}$, if $C$ is the conjugacy class of $g$, then $C = C^{-1}$ is real.

Suppose conversely that $G$ contains a non-trivial real conjugacy class $C$. Let $g \in C$ and let $N_G(g) = \{x \in G \mid xg = gx\}$ be the normalizer of $g$. Then $|C| = [G : N_G(g)]$. Suppose that $hgh^{-1} = g^{-1}$. Then

$$h^2gh^{-2} = hg^{-1}h^{-1} = (hgh^{-1})^{-1} = g$$

and so $h^2 \in N_G(g)$. If $h \in \langle h^2 \rangle$, then $h \in N_G(g)$ and so $g^{-1} = hgh^{-1} = g$. Hence in this case $g^2 = 1$ and so $|G|$ is even. If $h \notin \langle h^2 \rangle$, then $h^2$ is not a generator of $\langle h \rangle$ and so 2 divides the order of $h$. Thus $|G|$ is even. This completes the proof. □

From Proposition 9.1.11, we deduce a curious result about groups of odd order that does not seem to admit a direct elementary proof.

**Theorem 9.1.12 (Burnside).** *Let $G$ be a group of odd order and let $s$ be the number of conjugacy classes of $G$. Then $s \equiv |G| \bmod 16$.*

*Proof.* By Proposition 9.1.11, $G$ has the trivial character $\chi_0$ and the remaining characters come in conjugate pairs $\chi_1, \chi_1', \ldots, \chi_k, \chi_k'$ of degrees $d_1, \ldots, d_k$. In particular, $s = 1 + 2k$ and

$$|G| = 1 + \sum_{j=1}^{k} 2d_j^2.$$

Since $d_j$ divides $|G|$ it is odd and so we may write it as $d_j = 2m_j + 1$ for some non-negative integer $m_j$. Therefore, we have

$$\begin{aligned} |G| &= 1 + \sum_{j=1}^{k} 2(2m_j + 1)^2 = 1 + \sum_{j=1}^{k} (8m_j^2 + 8m_j + 2) \\ &= 1 + 2k + 8\sum_{j=1}^{k} m_j(m_j + 1) = s + 8\sum_{j=1}^{k} m_j(m_j + 1) \\ &\equiv s \bmod 16 \end{aligned}$$

since exactly one of $m_j$ and $m_j + 1$ is even. □

## Exercises

**Exercise 9.1.** Let $G$ be a finite group.

1. Prove that two elements $g, h \in G$ are conjugate if and only if $\chi(g) = \chi(h)$ for all irreducible characters $\chi$.
2. Show that the conjugacy class $C$ of an element $g \in G$ is real if and only if $\chi(g)$ is real for all irreducible characters $\chi$.

**Exercise 9.2.** Prove that all characters of the symmetric group $S_n$ are real.

**Exercise 9.3.** Let $G$ be a non-abelian group of order 155. Prove that $G$ has 11 conjugacy classes.

**Exercise 9.4.** Let $G$ be a group and let $\alpha$ be an automorphism of $G$.

1. If $\varphi\colon G \longrightarrow GL(V)$ is a representation, show that $\alpha^*(\varphi) = \varphi \circ \alpha$ is a representation of $G$.
2. If $f \in L(G)$, define $\alpha^*(f) = f \circ \alpha$. Prove that $\chi_{\alpha^*(\varphi)} = \alpha^*(\chi_\varphi)$.
3. Prove that if $\varphi$ is irreducible, then so is $\alpha^*(\varphi)$.
4. Show that if $C$ is a conjugacy class of $G$, then $\alpha(C)$ is also a conjugacy class of $G$.
5. Prove that the number of conjugacy classes $C$ with $\alpha(C) = C$ is equal to the number of irreducible characters $\chi$ with $\alpha^*(\chi) = \chi$. (Hint: imitate the proof of Theorem 9.1.10; the key point is that permuting the rows of the character table according to $\alpha^*$ yields the same matrix as permuting the columns according to $\alpha$.)

# Chapter 10
# Representation Theory of the Symmetric Group

No group is of greater importance than the symmetric group. After all, any group can be embedded as a subgroup of a symmetric group. In this chapter, we construct the irreducible representations of the symmetric group $S_n$. The character theory of the symmetric group is a rich and important theory filled with important connections to combinatorics. One can find whole books dedicated to this subject, cf. [12, 16, 17, 19]. Moreover, there are important applications to such diverse areas as voting and card shuffling [3, 7, 8].

## 10.1 Partitions and Tableaux

We begin with the fundamental notion of a partition of $n$. Simply speaking, a partition of $n$ is a way of writing $n$ as a sum of positive integers.

**Definition 10.1.1 (Partition).** A *partition of* $n$ is a tuple $\lambda = (\lambda_1, \ldots, \lambda_\ell)$ of positive integers such that $\lambda_1 \geq \lambda_2 \geq \cdots \geq \lambda_\ell$ and $\lambda_1 + \cdots + \lambda_\ell = n$. To indicate that $\lambda$ is a partition of $n$, we write $\lambda \vdash n$.

For example, $(2, 2, 1, 1)$ is partition of 6 and $(3, 1)$ is partition of 4. Note that $(1, 2, 1)$ is not a partition of 4 since the second entry is larger than the first.

There is a natural partition of $n$ associated to any permutation $\sigma \in S_n$ called the *cycle type* of $\sigma$. Namely, $\mathrm{type}(\sigma) = (\lambda_1, \ldots, \lambda_\ell)$ where the $\lambda_i$ are the lengths of the cycles of $\sigma$ in decreasing order (with multiplicity). Here we must count cycles of length 1, which are normally omitted from the notation when writing cycle decompositions.

*Example 10.1.2.* Let $n = 5$. Then

$$\mathrm{type}((1\ 2)(5\ 3\ 4)) = (3, 2)$$

$$\mathrm{type}((1\ 2\ 3)) = (3, 1, 1)$$

B. Steinberg, *Representation Theory of Finite Groups: An Introductory Approach*, Universitext, DOI 10.1007/978-1-4614-0776-8_10,
© Springer Science+Business Media, LLC 2012

$$\text{type}((1\ 2\ 3\ 4\ 5)) = (5)$$

$$\text{type}((1\ 2)(3\ 4)) = (2, 2, 1).$$

It is typically shown in a first course in group theory that two permutations are conjugate if and only if they have the same cycle type.

**Theorem 10.1.3.** *Let $\sigma, \tau \in S_n$. Then $\sigma$ is conjugate to $\tau$ if and only if* $\text{type}(\sigma) = \text{type}(\tau)$.

It follows that the number of irreducible representations of $S_n$ is the number of partitions of $n$. Thus we expect partitions to play a major role in the representation theory of the symmetric group. Our goal in this chapter is to give an explicit bijection between partitions of $n$ and irreducible representations of $S_n$. First we need to deal with some preliminary combinatorics.

It is often convenient to represent partitions by a "Tetris-like" picture called a Young diagram.

**Definition 10.1.4 (Young diagram).** If $\lambda = (\lambda_1, \ldots, \lambda_\ell)$ is a partition of $n$, then the *Young diagram* (or simply *diagram*) of $\lambda$ consists of $n$ boxes placed into $\ell$ rows where the $i$th row has $\lambda_i$ boxes.

This definition is best illustrated with an example. If $\lambda = (3, 1)$, then the Young diagram is as follows.

(10.1)

Conversely, any diagram consisting of $n$ boxes arranged into rows such that the number of boxes in each row is non-increasing (going from top to bottom) is the Young diagram of a unique partition of $n$.

**Definition 10.1.5 (Conjugate partition).** If $\lambda \vdash n$, then the *conjugate partition* $\lambda^T$ of $\lambda$ is the partition whose Young diagram is the transpose of the diagram of $\lambda$, that is, the Young diagram of $\lambda^T$ is obtained from the diagram of $\lambda$ by exchanging rows and columns.

Again, a picture is worth one thousand words.

*Example 10.1.6.* If $\lambda = (3, 1)$, then its diagram is as in (10.1). The transpose diagram is

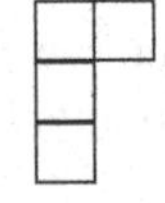

and so $\lambda^T = (2, 1, 1)$.

Next, we want to introduce an ordering on partitions. Given two partitions $\lambda$ and $\mu$ of $n$, we want to say that $\lambda$ dominates $\mu$, written $\lambda \trianglerighteq \mu$, if, for every $i \geq 1$, the first $i$ rows of the diagram of $\lambda$ contain at least as many boxes as the first $i$ rows of $\mu$.

*Example 10.1.7.* For instance, $(5,1) \trianglerighteq (3,3)$ as we can see from

$(5,1) =$ 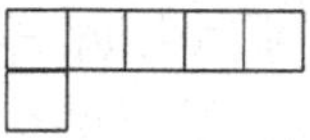 and $(3,3) =$ 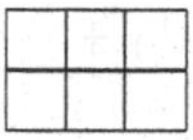

On the other hand, neither $(3,3,1) \trianglerighteq (4,1,1,1)$, nor $(4,1,1,1) \trianglerighteq (3,3,1)$ holds because $(4,1,1,1)$ has more elements in the first row, but $(3,3,1)$ has more elements in the first two rows.

$(3,3,1) =$ 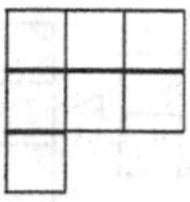 $(4,1,1,1) =$ 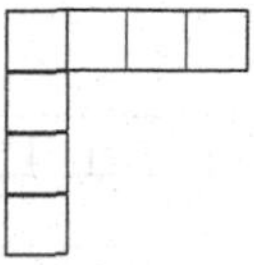

Let us formalize the definition. Observe that the number of boxes in the first $i$ rows of $\lambda = (\lambda_1, \ldots, \lambda_\ell)$ is $\lambda_1 + \cdots + \lambda_i$.

**Definition 10.1.8 (Domination order).** Suppose that $\lambda = (\lambda_1, \ldots, \lambda_\ell)$ and $\mu = (\mu_1, \ldots, \mu_m)$ are partitions of $n$. Then $\lambda$ is said to *dominate* $\mu$ if

$$\lambda_1 + \lambda_2 + \cdots + \lambda_i \geq \mu_1 + \mu_2 + \cdots + \mu_i$$

for all $i \geq 1$ where if $i > \ell$, then we take $\lambda_i = 0$, and if $i > m$, then we take $\mu_i = 0$.

The domination order satisfies many of the properties enjoyed by $\geq$.

**Proposition 10.1.9.** *The dominance order satisfies:*

1. *Reflexivity:* $\lambda \trianglerighteq \lambda$;
2. *Anti-symmetry:* $\lambda \trianglerighteq \mu$ *and* $\mu \trianglerighteq \lambda$ *implies* $\lambda = \mu$;
3. *Transitivity:* $\lambda \trianglerighteq \mu$ *and* $\mu \trianglerighteq \rho$ *implies* $\lambda \trianglerighteq \rho$.

*Proof.* Reflexivity is clear. Suppose $\lambda = (\lambda_1, \ldots, \lambda_\ell)$ and $\mu = (\mu_1, \ldots, \mu_m)$ are partitions of $n$. We prove by induction on $n$ that $\lambda \trianglerighteq \mu$ and $\mu \trianglerighteq \lambda$ implies $\lambda = \mu$. If $n = 1$, then $\lambda = (1) = \mu$ and there is nothing to prove. Otherwise, by taking $i = 1$, we see that $\lambda_1 = \mu_1$. Call this common value $k > 0$. Then define partitions $\lambda'$, $\mu'$ of $n - k$ by $\lambda' = (\lambda_2, \ldots, \lambda_\ell)$ and $\mu' = (\mu_2, \ldots, \mu_m)$. Since

$$\lambda_1 + \lambda_2 + \cdots + \lambda_i = \mu_1 + \mu_2 + \cdots + \mu_i$$

for all $i \geq 1$ and $\lambda_1 = \mu_1$, it follows that

$$\lambda_2 + \cdots + \lambda_i = \mu_2 + \cdots + \mu_i$$

for all $i \geq 1$ and hence $\lambda' \trianglerighteq \mu'$ and $\mu' \trianglerighteq \lambda'$. Thus by induction $\lambda' = \mu'$ and hence $\lambda = \mu$. This establishes anti-symmetry.

To obtain transitivity, simply observe that

$$\lambda_1 + \cdots + \lambda_i \geq \mu_1 + \cdots + \mu_i \geq \rho_1 + \cdots + \rho_i$$

and so $\lambda \trianglerighteq \rho$. □

Proposition 10.1.9 says that $\trianglerighteq$ is a *partial order* on the set of partitions of $n$.

*Example 10.1.10.*

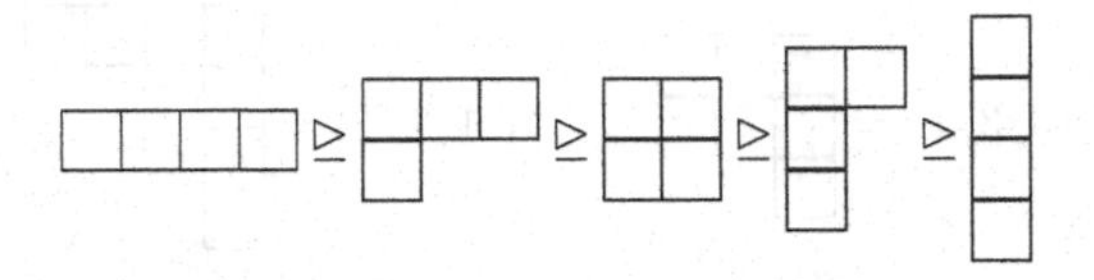

Young tableaux are obtained from Young diagrams by placing the integers $1, \ldots, n$ into the boxes.

**Definition 10.1.11 (Young tableaux).** If $\lambda \vdash n$, then a $\lambda$*-tableaux* (or *Young tableaux of shape* $\lambda$) is an array $t$ of integers obtained by placing $1, \ldots, n$ into the boxes of the Young diagram for $\lambda$. There are clearly $n!$ $\lambda$-tableaux.

This concept is again best illustrated with an example.

*Example 10.1.12.* Suppose that $\lambda = (3, 2, 1)$. Then some $\lambda$-tableaux are as follows.

$$\begin{array}{|c|c|c|}\hline 1 & 2 & 3 \\ \hline 4 & 5 \\ \cline{1-2} 6 \\ \cline{1-1}\end{array}\,, \quad \begin{array}{|c|c|c|}\hline 3 & 1 & 2 \\ \hline 6 & 5 \\ \cline{1-2} 4 \\ \cline{1-1}\end{array}\,, \quad \begin{array}{|c|c|c|}\hline 3 & 2 & 6 \\ \hline 1 & 5 \\ \cline{1-2} 4 \\ \cline{1-1}\end{array}$$

A rather technical combinatorial fact is that if $t^\lambda$ is a $\lambda$-tableaux and $s^\mu$ is a $\mu$-tableaux such that the integers in any given row of $s^\mu$ belong to distinct columns of $t^\lambda$, then $\lambda \trianglerighteq \mu$. To prove this, we need the following proposition, which will be useful in its own right.

**Proposition 10.1.13.** *Let $\lambda = (\lambda_1, \ldots, \lambda_\ell)$ and $\mu = (\mu_1, \ldots, \mu_m)$ be partitions of $n$. Suppose that $t^\lambda$ is a $\lambda$-tableaux and $s^\mu$ is a $\mu$-tableaux such that entries in the same row of $s^\mu$ are located in different columns of $t^\lambda$. Then we can find a $\lambda$-tableaux $u^\lambda$ such that:*

1. *The $j$th columns of $t^\lambda$ and $u^\lambda$ contain the same elements for $1 \leq j \leq \ell$;*
2. *The entries of the first $i$ rows of $s^\mu$ belong to the first $i$ rows of $u^\lambda$ for each $1 \leq i \leq m$.*

*Proof.* For each $1 \leq r \leq m$, we construct a $\lambda$-tableaux $t^\lambda_r$ such that:

(a) The $j$th columns of $t^\lambda$ and $t^\lambda_r$ contain the same elements for $1 \leq j \leq \ell$;
(b) The entries of the first $i$ rows of $s^\mu$ belong to the first $i$ rows of $t^\lambda_r$ for $1 \leq i \leq r$.

Setting $u^\lambda = t^\lambda_m$ will then complete the proof. The construction is by induction on $r$. Let us begin with $r = 1$. Let $k$ be an element in the first row of $s^\mu$ and let $c(k)$ be the column of $t^\lambda$ containing $k$. If $k$ is in the first row of $t^\lambda$, we do nothing. Otherwise, we switch in $t^\lambda$ the first entry in $c(k)$ with $k$. Because each element of the first row of $s^\mu$ is in a different column of $t^\lambda$, the order in which we do this does not matter, and so we can construct a new $\lambda$-tableaux $t^\lambda_1$ satisfying properties (a) and (b).

Next suppose that $t^\lambda_r$ with the desired two properties has been constructed for $r$ where $1 \leq r \leq m-1$. Define $t^\lambda_{r+1}$ as follows. Let $k$ be an entry of row $r+1$ of $s^\mu$ and let $c(k)$ be the column in which $k$ appears in $t^\lambda_r$. If $k$ already appears in the first $r+1$ rows of $t^\lambda_r$, there is nothing to do. So assume that $k$ does not appear in the first $r+1$ rows of $t^\lambda_r$. Notice that if row $r+1$ of $t^\lambda_r$ does not intersect $c(k)$, then because the sizes of the rows are non-increasing, it follows that $k$ already appears in the first $r$ rows of $t^\lambda_r$. Thus we must have that $c(k)$ intersects row $r+1$. Thus we can switch $k$ with the element in row $r+1$ and column $c(k)$ of $t^\lambda_r$ without losing property (b) for $1 \leq i \leq r$. Again, because each entry of row $r+1$ of $s^\mu$ is in a different column of $t^\lambda$, and hence of $t^\lambda_r$ by property (a), we can do this for each such $k$ independently. In this way, we have constructed $t^\lambda_{r+1}$ satisfying (a) and (b). □

Let us illustrate how this works with an example.

*Example 10.1.14.* Suppose that $t^\lambda$ and $s^\mu$ are given by

$$t^\lambda = \begin{array}{|c|c|c|c|c|}\hline 8&5&4&2&7\\\hline 1&3\\\cline{1-2} 6\\\cline{1-1}\end{array} \quad \text{and} \quad s^\mu = \begin{array}{|c|c|c|c|}\hline 1&2&3&4\\\hline 5&6\\\cline{1-2} 7&8\\\cline{1-2}\end{array}.$$

No two elements in the same row of $s^\mu$ belong to the same column of $t^\lambda$.

We construct $t^\lambda_1$ by switching in $t^\lambda$ each element appearing of the first row of $s^\mu$ with the element in its column of the first row of $t^\lambda$, so

$$t^\lambda_1 = \begin{array}{|c|c|c|c|c|}\hline 1&3&4&2&7\\\hline 8&5\\\cline{1-2} 6\\\cline{1-1}\end{array}.$$

Now by switching 8 and 6, we obtain the $\lambda$-tableaux

$$t^\lambda_2 = \begin{array}{|c|c|c|c|c|}\hline 1&3&4&2&7\\\hline 6&5\\\cline{1-2} 8\\\cline{1-1}\end{array}$$

which has every element in the first $i$ rows of $s^\mu$ located in the first $i$ rows of $t^\lambda_2$ for $i = 1, 2, 3$. Hence we can take $u^\lambda = t^\lambda_2$.

Our first use of Proposition 10.1.13 is to establish the following combinatorial criterion for domination.

**Lemma 10.1.15 (Dominance lemma).** *Let $\lambda$ and $\mu$ be partitions of $n$ and suppose that $t^\lambda$ and $s^\mu$ are tableaux of respective shapes $\lambda$ and $\mu$. Moreover, suppose that integers in the same row of $s^\mu$ are located in different columns of $t^\lambda$. Then $\lambda \trianglerighteq \mu$.*

*Proof.* Let $\lambda = (\lambda_1, \ldots, \lambda_\ell)$ and $\mu = (\mu_1, \ldots, \mu_m)$. By Proposition 10.1.13 we can find a $\lambda$-tableaux $u^\lambda$ such that, for $1 \leq i \leq m$, the entries of the first $i$ rows of $s^\mu$ are in the first $i$ rows of $u^\lambda$. Then since $\lambda_1 + \cdots + \lambda_i$ is the number of entries in the first $i$ rows of $u^\lambda$ and $\mu_1 + \cdots + \mu_i$ is the number of entries in the first $i$ rows of $s^\mu$, it follows that $\lambda_1 + \cdots + \lambda_i \geq \mu_1 + \cdots + \mu_i$ for all $i \geq 1$ and hence $\lambda \trianglerighteq \mu$. □

## 10.2 Constructing the Irreducible Representations

If $X \subseteq \{1, \ldots, n\}$, we identify $S_X$ with those permutations in $S_n$ that fix all elements outside of $X$. For instance, $S_{\{2,3\}}$ consists of $\{Id, (2\ 3)\}$.

**Definition 10.2.1 (Column stabilizer).** Let $t$ be a Young tableaux. Then the *column stabilizer* of $t$ is the subgroup of $S_n$ preserving the columns of $t$. That is, $\sigma \in C_t$ if and only if $\sigma(i)$ is in the same column as $i$ for each $i \in \{1, \ldots, n\}$.

Let us turn to an example.

*Example 10.2.2.* Suppose that

$$t = \begin{array}{|c|c|c|}\hline 1 & 3 & 7 \\ \hline 4 & 5 \\ \cline{1-2} 2 & 6 \\ \cline{1-2}\end{array} .$$

Then $C_t = S_{\{1,2,4\}}S_{\{3,5,6\}}S_{\{7\}} \cong S_{\{1,2,4\}} \times S_{\{3,5,6\}} \times S_{\{7\}}$. So, for example, $(1\ 4), (1\ 2\ 4)(3\ 5) \in C_t$. Since $S_{\{7\}} = \{Id\}$, it follows $|C_t| = 3! \cdot 3! = 36$.

The group $S_n$ acts transitively on the set of $\lambda$-tableaux by applying $\sigma \in S_n$ to the entries of the boxes. The result of applying $\sigma \in S_n$ to $t$ is denoted $\sigma t$. For example, if

$$t = \begin{array}{|c|c|c|}\hline 1 & 3 & 4 \\ \hline 2 \\ \cline{1-1}\end{array}$$

and $\sigma = (1\ 3\ 2)$, then

$$\sigma t = \begin{array}{|c|c|c|}\hline 3 & 2 & 4 \\ \hline 1 \\ \cline{1-1}\end{array} .$$

Let us define an equivalence relation $\sim$ on the set of $\lambda$-tableaux by putting $t_1 \sim t_2$ if they have the same entries in each row. For example,

$$\begin{array}{|c|c|c|}\hline 1 & 2 & 3 \\ \hline 4 & 5 \\ \cline{1-2}\end{array} \sim \begin{array}{|c|c|c|}\hline 3 & 1 & 2 \\ \hline 5 & 4 \\ \cline{1-2}\end{array}$$

since they both have $\{1, 2, 3\}$ in the first row and $\{4, 5\}$ in the second row.

**Definition 10.2.3 (Tabloid).** A $\sim$-equivalence class of $\lambda$-tableaux is called a $\lambda$-*tabloid* or a *tabloid of shape* $\lambda$. The tabloid of a tableaux $t$ is denoted $[t]$. The set of all tabloids of shape $\lambda$ is denoted $T^\lambda$. Denote by $T_\lambda$ the tabloid with $1, \ldots, \lambda_1$ in row 1, $\lambda_1 + 1, \ldots, \lambda_1 + \lambda_2$ in row 2 and in general with $\lambda_1 + \cdots + \lambda_{i-1} + 1, \ldots, \lambda_1 + \cdots + \lambda_i$ in row $i$. In other words, $T_\lambda$ is the tabloid corresponding to the tableaux which has $j$ in the $j$th box.

For example, $T_{(3,2)}$ is the equivalence class of

| 1 | 2 | 3 |
|---|---|---|
| 4 | 5 | |

.

Our next proposition implies that the action of $S_n$ on $\lambda$-tableaux induces a well-defined action of $S_n$ on tabloids of shape $\lambda$.

**Proposition 10.2.4.** *Suppose that $t_1 \sim t_2$ and $\sigma \in S_n$. Then $\sigma t_1 \sim \sigma t_2$. Hence there is a well-defined action of $S_n$ on $T^\lambda$ given by putting $\sigma[t] = [\sigma t]$ for $t$ a $\lambda$-tableaux.*

*Proof.* To show that $\sigma t_1 \sim \sigma t_2$, we must show that $i, j$ are in the same row of $\sigma t_1$ if and only if they are in the same row of $\sigma t_2$. But $i, j$ are in the same row of $\sigma t_1$ if and only if $\sigma^{-1}(i)$ and $\sigma^{-1}(j)$ are in the same row of $t_1$, which occurs if and only if $\sigma^{-1}(i)$ and $\sigma^{-1}(j)$ are in the same row of $t_2$. But this occurs if and only if $i, j$ are in the same row of $\sigma t_2$. This proves that $\sigma t_1 \sim \sigma t_2$. From this it is easy to verify that $\sigma[t] = [\sigma t]$ gives a well-defined action of $S_n$ on $T^\lambda$. □

The action of $S_n$ on $\lambda$-tabloids is transitive since it was already transitive on $\lambda$-tableaux. Suppose that $\lambda = (\lambda_1, \ldots, \lambda_\ell)$. The stabilizer $S_\lambda$ of $T_\lambda$ is

$$S_\lambda = S_{\{1,\ldots,\lambda_1\}} \times S_{\{\lambda_1+1,\ldots,\lambda_1+\lambda_2\}} \times \cdots \times S_{\{\lambda_1+\cdots+\lambda_{\ell-1}+1,\ldots,n\}}.$$

Thus $|T^\lambda| = [S_n : S_\lambda] = n!/\lambda_1! \cdots \lambda_\ell!$. The subgroup $S_\lambda$ is called the *Young subgroup* associated to the partition $\lambda$.

For a partition $\lambda$, set $M^\lambda = \mathbb{C}T^\lambda$ and let $\varphi^\lambda \colon S_n \longrightarrow GL(M^\lambda)$ be the associated permutation representation.

*Example 10.2.5.* Suppose that $\lambda = (n-1, 1)$. Then two $\lambda$-tableaux are equivalent if and only if they have the same entry in the second row. Thus $T^\lambda$ is in bijection with $\{1, \ldots, n\}$ and $\varphi^\lambda$ is equivalent to the standard representation. On the other hand, if $\lambda = (n)$, then there is only one $\lambda$-tabloid and so $\varphi^\lambda$ is the trivial representation.

If $\lambda = (n-2, 2)$, then two $\lambda$-tabloids are equivalent if and only if the two-element subsets in the second row of the tabloids coincide. Hence $\lambda$-tabloids are in bijection with two-element subsets of $\{1, \ldots, n\}$ and $\varphi^\lambda$ is equivalent to the permutation representation associated to the action of $S_n$ on $[n]^2$. In particular, $(S_n, S_{(n-2,2)})$ is a Gelfand pair.

If $\lambda \neq (n)$, then $\varphi^\lambda$ is a non-trivial permutation representation of $S_n$ and hence is not irreducible. Nonetheless, it contains a distinguished irreducible constituent that we now seek to isolate.

**Definition 10.2.6 (Polytabloid).** Let $\lambda, \mu \vdash n$. Let $t$ be a $\lambda$-tableaux and define a linear operator $A_t\colon M^\mu \longrightarrow M^\mu$ by

$$A_t = \sum_{\pi \in C_t} \mathrm{sgn}(\pi)\varphi^\mu_\pi.$$

In the case $\lambda = \mu$, the element

$$e_t = A_t[t] = \sum_{\pi \in C_t} \mathrm{sgn}(\pi)\pi[t]$$

of $M^\lambda$ is called the *polytabloid* associated to $t$.

Our next proposition shows that the action of $S_n$ on $\lambda$-tableaux is compatible with the definition of a $\lambda$-tabloid.

**Proposition 10.2.7.** *If $\sigma \in S_n$ and $t$ is a $\lambda$-tableaux, then $\varphi^\lambda_\sigma e_t = e_{\sigma t}$.*

*Proof.* First we claim that $C_{\sigma t} = \sigma C_t \sigma^{-1}$. Indeed, if $X_i$ is the set of entries of column $i$ of $t$, then $\sigma(X_i)$ is the set of entries of column $i$ of $\sigma t$. Since $\tau$ stabilizes $X_i$ if and only if $\sigma\tau\sigma^{-1}$ stabilizes $\sigma(X_i)$, the claim follows. Now we compute

$$\begin{aligned}
\varphi^\lambda_\sigma A_t &= \sum_{\pi \in C_t} \mathrm{sgn}(\pi)\varphi^\lambda_\sigma\varphi^\lambda_\pi \\
&= \sum_{\tau \in C_{\sigma t}} \mathrm{sgn}(\sigma^{-1}\tau\sigma)\varphi^\lambda_\sigma\varphi^\lambda_{\sigma^{-1}\tau\sigma} \\
&= A_{\sigma t}\varphi^\lambda_\sigma
\end{aligned}$$

where we have made the substitution $\tau = \sigma\pi\sigma^{-1}$.

Thus $\varphi^\lambda_\sigma e_t = \varphi^\lambda_\sigma A_t[t] = A_{\sigma t}\varphi^\lambda_\sigma[t] = A_{\sigma t}[\sigma t] = e_{\sigma t}$. This completes the proof. □

We can now define our desired subrepresentation.

**Definition 10.2.8 (Sprecht representation).** Let $\lambda$ be a partition of $n$. Define $S^\lambda$ to be the subspace of $M^\lambda$ spanned by the polytabloids $e_t$ with $t$ a $\lambda$-tableaux. Proposition 10.2.7 implies that $S^\lambda$ is $S_n$-invariant. Let $\psi^\lambda\colon S_n \longrightarrow GL(S^\lambda)$ be the corresponding subrepresentation. It is called the *Sprecht representation* associated to $\lambda$.

*Remark 10.2.9.* The $e_t$ are not in general linearly independent. See the next example.

Our goal is to prove that the Sprecht representations $\psi^\lambda$ form a complete set of irreducible representations of $S_n$. Let us look at an example.

*Example 10.2.10 (Alternating representation).* Consider the partition $\lambda = (1, 1, \ldots, 1)$ of $n$. Since each row has only one element, $\lambda$-tableaux are the same thing as $\lambda$-tabloids. Thus $\varphi^\lambda$ is equivalent to the regular representation of $S_n$. Let $t$ be a $\lambda$-tableaux. Because $t$ has only one column, trivially $C_t = S_n$. Thus

$$e_t = \sum_{\pi \in S_n} \operatorname{sgn}(\pi)\pi[t].$$

We claim that if $\sigma \in S_n$, then $\varphi^\lambda_\sigma e_t = \operatorname{sgn}(\sigma)e_t$. Since we know that $\varphi^\lambda_\sigma e_t = e_{\sigma t}$ by Proposition 10.2.7, it will follow that $S^\lambda = \mathbb{C}e_t$ and that $\psi^\lambda$ is equivalent to the degree one representation $\operatorname{sgn}\colon S_n \longrightarrow \mathbb{C}^*$.

Indeed, we compute

$$\begin{aligned}\varphi^\lambda_\sigma e_t &= \sum_{\pi \in S_n} \operatorname{sgn}(\pi)\varphi^\lambda_\sigma \varphi^\lambda_\pi [t] \\ &= \sum_{\tau \in S_n} \operatorname{sgn}(\sigma^{-1}\tau)\varphi^\lambda_\tau [t] \\ &= \operatorname{sgn}(\sigma)e_t\end{aligned}$$

where we have performed the substitution $\tau = \sigma\pi$.

The proof that the $\psi^\lambda$ are the irreducible representations of $S_n$ proceeds via a series of lemmas.

**Lemma 10.2.11.** *Let $\lambda, \mu \vdash n$ and suppose that $t^\lambda$ is a $\lambda$-tableaux and $s^\mu$ is a $\mu$-tableaux such that $A_{t^\lambda}[s^\mu] \neq 0$. Then $\lambda \trianglerighteq \mu$. Moreover, if $\lambda = \mu$, then $A_{t^\lambda}[s^\mu] = \pm e_{t^\lambda}$.*

*Proof.* We use the dominance lemma. Suppose that we have two elements $i, j$ that are in the same row of $s^\mu$ and the same column of $t^\lambda$. Then $(i\ j)[s^\mu] = [s^\mu] - Id[s^\mu]$ and thus

$$(\varphi^\mu_{Id} - \varphi^\mu_{(i\ j)})[s^\mu] = 0. \tag{10.2}$$

Let $H = \{Id, (i\ j)\}$. Then $H$ is a subgroup of $C_{t^\lambda}$. Let $\sigma_1, \ldots, \sigma_k$ be a complete set of left coset representatives for $H$ in $C_{t^\lambda}$. Then we have

$$\begin{aligned}A_{t^\lambda}[s^\mu] &= \sum_{\pi \in C_{t^\lambda}} \operatorname{sgn}(\pi)\varphi^\mu_\pi [s^\mu] \\ &= \sum_{r=1}^{k} \left(\operatorname{sgn}(\sigma_r)\varphi^\mu_{\sigma_r} + \operatorname{sgn}(\sigma_r(i\ j))\varphi^\mu_{\sigma_r(i\ j)}\right)[s^\mu] \\ &= \sum_{r=1}^{k} \operatorname{sgn}(\sigma_r)\varphi^\mu_{\sigma_r}(\varphi^\mu_{Id} - \varphi^\mu_{(i\ j)})[s^\mu] \\ &= 0\end{aligned}$$

where the last equality uses (10.2). This contradiction implies that the elements of each row of $s^\mu$ are in different columns of $t^\lambda$. The dominance lemma (Lemma 10.1.15) now yields that $\lambda \trianglerighteq \mu$.

Next suppose that $\lambda = \mu$. Let $u^\lambda$ be as in Proposition 10.1.13. The fact that the columns of $t^\lambda$ and $u^\lambda$ have the same elements implies that the unique permutation $\sigma$ with $u^\lambda = \sigma t^\lambda$ belongs to $C_{t^\lambda}$. On the other hand, for all $i \geq 1$, the first $i$ rows of $s^\mu$ belong to the first $i$ rows of $u^\lambda$. But since $\lambda = \mu$, this implies $[u^\lambda] = [s^\mu]$. Indeed, the first row of $s^\mu$ is contained in the first row of $u^\lambda$. But both rows have the same number of boxes. Therefore, these rows must contain the same elements. Suppose by induction, that each of the first $i$ rows of $u^\lambda$ and $s^\mu$ have the same elements. Then since each element of the first $i+1$ rows of $s^\mu$ belongs to the first $i+1$ rows of $u^\lambda$, it follows from the inductive hypothesis that each element of row $i+1$ of $s^\mu$ belongs to row $i+1$ of $u^\lambda$. Since these tableaux both have shape $\lambda$, it follows that they have the same $(i+1)^{st}$ row. We conclude that $[u^\lambda] = [s^\mu]$.

It follows that

$$\begin{aligned} A_{t^\lambda}[s^\mu] &= \sum_{\pi \in C_{t^\lambda}} \operatorname{sgn}(\pi)\varphi^\lambda_\pi[s^\mu] \\ &= \sum_{\tau \in C_{t^\lambda}} \operatorname{sgn}(\tau\sigma^{-1})\varphi^\lambda_\tau\varphi^\lambda_{\sigma^{-1}}[u^\lambda] \\ &= \operatorname{sgn}(\sigma^{-1}) \sum_{\tau \in C_{t^\lambda}} \operatorname{sgn}(\tau)\tau[t^\lambda] \\ &= \pm e_{t^\lambda} \end{aligned}$$

where in the second equality we have performed the change of variables $\tau = \pi\sigma$. This completes the proof. □

The next lemma continues our study of the operator $A_t$.

**Lemma 10.2.12.** *Let $t$ be a $\lambda$-tableaux. Then the image of the operator $A_t : M^\lambda \longrightarrow M^\lambda$ is $\mathbb{C}e_t$.*

*Proof.* From the equation $e_t = A_t[t]$, it suffices to show that the image is contained in $\mathbb{C}e_t$. To prove this, it suffices to check on basis elements $[s] \in T^\lambda$. If $A_t[s] = 0$, there is nothing to prove; otherwise, Lemma 10.2.11 yields $A_t[s] = \pm e_t \in \mathbb{C}e_t$. This completes the proof. □

Recall that $M^\lambda = \mathbb{C}T^\lambda$ comes equipped with an inner product for which $T^\lambda$ is an orthonormal basis and that, moreover, the representation $\varphi^\lambda$ is unitary with respect to this product. Furthermore, if $t$ is a $\lambda$-tableaux, then

$$A_t^* = \sum_{\pi \in C_t} \operatorname{sgn}(\pi)(\varphi^\lambda_\pi)^* = \sum_{\tau \in C_t} \operatorname{sgn}(\tau)\varphi^\lambda_\tau = A_t$$

where the penultimate equality is obtained by setting $\tau = \pi^{-1}$ and using that $\varphi$ is unitary. Thus $A_t$ is self-adjoint.

The key to proving that the $\psi^\lambda$ are the irreducible representations of $S_n$ is the following theorem.

**Theorem 10.2.13 (Subrepresentation theorem).** *Let $\lambda$ be a partition of $n$ and suppose that $V$ is an $S_n$-invariant subspace of $M^\lambda$. Then either $S^\lambda \subseteq V$ or $V \subseteq (S^\lambda)^\perp$.*

*Proof.* Suppose first that there is a $\lambda$-tableaux $t$ and a vector $v \in V$ such that $A_t v \neq 0$. Then by Lemma 10.2.12 and $S_n$-invariance of $V$, we have $0 \neq A_t v \in \mathbb{C}e_t \cap V$. It follows that $e_t \in V$. Hence, for all $\sigma \in S_n$, we have $e_{\sigma t} = \varphi^\lambda_\sigma e_t \in V$. Because $S_n$ acts transitively on the set of $\lambda$-tableaux, we conclude that $S^\lambda \subseteq V$.

Suppose next that, for all $\lambda$-tableaux $t$ and all $v \in V$, one has $A_t v = 0$. Then we have

$$\langle v, e_t \rangle = \langle v, A_t[t] \rangle = \langle A_t^* v, [t] \rangle = \langle A_t v, [t] \rangle = 0$$

because $A_t^* = A_t$ and $A_t v = 0$. As $t$ and $v$ were arbitrary, this shows that $V \subseteq (S^\lambda)^\perp$, completing the proof. □

As a corollary we see that $S^\lambda$ is irreducible.

**Corollary 10.2.14.** *Let $\lambda \vdash n$. Then $\psi^\lambda \colon S_n \longrightarrow GL(S^\lambda)$ is irreducible.*

*Proof.* Let $V$ be a proper $S_n$-invariant subspace of $S^\lambda$. Then by Theorem 10.2.13, we have $V \subseteq (S^\lambda)^\perp \cap S^\lambda = \{0\}$. This yields the corollary. □

We have thus constructed, for each partition $\lambda$ of $n$, an irreducible representation of $S_n$. The number of conjugacy classes of $S_n$ is the number of partitions of $n$. Hence, if we can show that $\lambda \neq \mu$ implies that $\psi^\lambda \nsim \psi^\mu$, then it will follow that we have found all the irreducible representations of $S_n$.

**Lemma 10.2.15.** *Suppose that $\lambda, \mu \vdash n$ and let $T \in \mathrm{Hom}_{S_n}(\varphi^\lambda, \varphi^\mu)$. If $S^\lambda \nsubseteq \ker T$, then $\lambda \trianglerighteq \mu$. Moreover, if $\lambda = \mu$, then $T|_{S^\lambda}$ is a scalar multiple of the identity map.*

*Proof.* Theorem 10.2.13 implies that $\ker T \subseteq (S^\lambda)^\perp$. So, for any $\lambda$-tableaux $t$, it follows that $0 \neq Te_t = TA_t[t] = A_t T[t]$, where the last equality uses that $T$ commutes with $\varphi^\lambda(S_n)$ and the definition of $A_t$. Now $T[t]$ is a linear combination of $\mu$-tabloids and so there exists a $\mu$-tabloid $[s]$ such that $A_t[s] \neq 0$. But then $\lambda \trianglerighteq \mu$ by Lemma 10.2.11.

Suppose now that $\lambda = \mu$. Then

$$Te_t = A_t T[t] \in \mathbb{C}e_t \subseteq S^\lambda$$

by Lemma 10.2.12. Thus $T$ leaves $S^\lambda$ invariant. Since $S^\lambda$ is irreducible, Schur's lemma implies $T|_{S^\lambda} = cI$ for some $c \in \mathbb{C}$. □

As a consequence we obtain the following result.

**Lemma 10.2.16.** *If* $\mathrm{Hom}_{S_n}(\psi^\lambda, \varphi^\mu) \neq 0$, *then* $\lambda \trianglerighteq \mu$. *Moreover, if* $\lambda = \mu$, *then* $\dim \mathrm{Hom}_{S_n}(\psi^\lambda, \varphi^\mu) = 1$.

*Proof.* Let $T\colon S^\lambda \longrightarrow M^\mu$ be a non-zero morphism of representations. Then we can extend $T$ to $M^\lambda = S^\lambda \oplus (S^\lambda)^\perp$ by putting $T(v+w) = Tv$ for elements $v \in S^\lambda$ and $w \in (S^\lambda)^\perp$. This extension is a morphism of representations because $(S^\lambda)^\perp$ is $S_n$-invariant and so

$$T(\varphi^\lambda_\sigma(v+w)) = T(\varphi^\lambda_\sigma v + \varphi^\lambda_\sigma w) = T\varphi^\lambda_\sigma v = \varphi^\mu_\sigma Tv = \varphi^\mu_\sigma T(v+w).$$

Clearly $S^\lambda \nsubseteq \ker T$ and so $\lambda \trianglerighteq \mu$ by Lemma 10.2.15. Moreover, if $\lambda = \mu$, then $T$ must be a scalar multiple of the inclusion map by Lemma 10.2.15 and so $\dim \mathrm{Hom}_{S_n}(\psi^\lambda, \varphi^\mu) = 1$. □

We can now prove the main result.

**Theorem 10.2.17.** *The Sprecht representations* $\psi^\lambda$ *with* $\lambda \vdash n$ *form a complete set of inequivalent irreducible representations of* $S_n$.

*Proof.* All that remains is to show that $\psi^\lambda \sim \psi^\mu$ implies $\lambda = \mu$. But $\psi^\lambda \sim \psi^\mu$, implies that $0 \neq \mathrm{Hom}_{S_n}(\psi^\lambda, \psi^\mu) \subseteq \mathrm{Hom}_{S_n}(\psi^\lambda, \varphi^\mu)$. Thus $\lambda \trianglerighteq \mu$ by Lemma 10.2.16. A symmetric argument shows that $\mu \trianglerighteq \lambda$ and so $\lambda = \mu$ by Proposition 10.1.9. This establishes the theorem. □

In fact, we can deduce more from Lemma 10.2.16.

**Corollary 10.2.18.** *Suppose* $\mu \vdash n$. *Then* $\psi^\mu$ *appears with multiplicity one as an irreducible constituent of* $\varphi^\mu$. *Any other irreducible constituent* $\psi^\lambda$ *of* $\varphi^\mu$ *satisfies* $\lambda \rhd \mu$.

## Exercises

**Exercise 10.1.** Verify that the relation $\sim$ on $\lambda$-tableaux is an equivalence relation.

**Exercise 10.2.** Verify that the action in Proposition 10.2.4 is indeed an action.

**Exercise 10.3.** Prove that if $\lambda = (n-1, 1)$, then the corresponding Sprecht representation of $S_n$ is equivalent to the augmentation subrepresentation of the standard representation of $S_n$.

**Exercise 10.4.** Compute the character table of $S_5$.

**Exercise 10.5.** Compute the decomposition of the representation $\varphi^{(3,2)}$ of $S_5$ into irreducible constituents.

**Exercise 10.6.** Prove Corollary 10.2.18.

**Exercise 10.7.** Prove that each character of $S_n$ takes on only rational values. (Hint: Observe that each polytabloid belongs to the $\mathbb{Q}$-span $\mathbb{Q}T^\lambda$ of the tabloids and that $\psi^\lambda$ leaves invariant the set of polytabloids. Deduce that if one chooses a basis $B$ for $S^\lambda$ contained in the set of polytabloids, then the matrices $[\psi^\lambda(\sigma)]_B$ with $\sigma \in S_n$ have rational entries.)

# Chapter 11
# Probability and Random Walks on Groups

One of the most important applications of group representation theory is to probability and statistics, via the study of random walks on groups. In a famous paper [1], Bayer and Diaconis gave very precise estimates on how many riffle shuffles it takes to randomize a deck of $n$ cards; the riffle shuffle is the shuffle where you cut the pack of cards into two halves and then interleave them. Bayer and Diaconis concluded based on their results that, for a deck of 52 cards, seven riffle shuffles are enough. Any fewer shuffles are too far from being random, whereas the net gain in randomness for doing more than seven shuffles is not sufficient to warrant the extra shuffles. However, it was later shown that there is a version of solitaire, called new age solitaire, for which seven shuffles is in fact too few. Bayer and Diaconis based their work on a model of riffle shuffling $n$ cards as a random walk on the symmetric group $S_n$. Properties of the algebra $L(S_n)$ play an important role in their analysis. A number of other card shuffling methods have also been analyzed using the representation theory of $S_n$ [7,9].

Random walks on finite groups are also good models of diffusion processes. A particularly well-known example is the Ehrenfest's urn model, which is a random walk on $(\mathbb{Z}/2\mathbb{Z})^n$. For further examples, see [3,7].

So what is a random walk, anyway? Imagine that a drunkard has just left the village pub and starts wandering off with no particular destination. We model the village with a graph where the vertices represent intersections and the edges represent streets. Each time the drunkard reaches an intersection, he randomly chooses a street and continues on his way. Natural questions include: What is the drunkard's probability of returning to the pub after $n$ steps? What is the probability distribution on the intersections that describes where the drunkard is after $n$ steps? How long does it take before he is equally likely to be at any intersection? If the graph is infinite, it is interesting to ask what is his probability of ever returning to the pub. If $\Gamma$ is the Cayley graph of a group $G$, then a random walk on $\Gamma$ is also referred to as a random walk on $G$.

Of course, studying drunkards ambling about does not seem all that worthy a task. However, there are many physical processes that are well modeled by random walks. For instance, our random walker could be a particle in a diffusion

B. Steinberg, *Representation Theory of Finite Groups: An Introductory Approach*, Universitext, DOI 10.1007/978-1-4614-0776-8_11,
© Springer Science+Business Media, LLC 2012

process. Or perhaps the vertices represent configurations of some objects, such as the orderings of a deck of cards, and the edges represent how one configuration can be transformed into another after a single step (e.g., how the deck can change after a single riffle shuffle). The random walk then models how one randomly goes from one configuration to the next.

## 11.1 Probabilities on Groups

Let $G$ be a finite group and suppose that $X$ is a $G$-valued *random variable*. Formally, this means that $X$ is a function $X\colon \Omega \longrightarrow G$, where $\Omega$ is some probability space (never mind what that means). The distribution of the random variable $X$ is the function $P\colon G \longrightarrow [0,1]$ defined by

$$P(g) = \mathrm{Prob}[X = g].$$

For all practical purposes, everything one needs to know about the random variable $X$ is encoded in its distribution and so one does not even need to know what is $\Omega$. Notice that $P$ satisfies

$$\sum_{g\in G} P(g) = 1. \tag{11.1}$$

Conversely, any function $P\colon G \longrightarrow [0,1]$ satisfying (11.1) is the distribution of some $G$-valued random variable. So in this book, we will work exclusively with probability distributions instead of random variables. This effectively means that the reader does not need to know any probability theory to read this chapter. Occasionally, we shall use the language of random variables to explain informally the intuition behind a notion.

**Definition 11.1.1 (Probability distribution).** A *probability distribution*, or simply a *probability*, on a finite group $G$ is a function $P\colon G \longrightarrow [0,1]$ such that (11.1) holds. If $A \subseteq G$, we put

$$P(A) = \sum_{g\in A} P(g).$$

The *support* of the probability $P$ is the set $\mathrm{supp}(P) = \{g \in G \mid P(g) \neq 0\}$.

Intuitively, if $P$ is a probability on $G$, then if we randomly choose an element $X$ from $G$ according to $P$, then the probability that $X = g$ is given by $P(g)$. More generally, if $A \subseteq G$, then $P(A)$ is the probability that an element $X$ of $G$ chosen at random according to $P$ belongs to $A$. For example, if $G = \mathbb{Z}/2\mathbb{Z}$ and $P([0]) = 1/2$, $P([1]) = 1/2$, then $P$ is a probability on $G$ for which $[0]$ and $[1]$ are equally probable. This can be generalized to any group.

**Definition 11.1.2 (Uniform distribution).** Let $G$ be a finite group. Then the *uniform distribution* $U$ on $G$ is given by

$$U(g) = \frac{1}{|G|}$$

for all $g \in G$.

One normally thinks of the uniform distribution as being unbiased. Usually, when someone informally speaks of choosing an element at random, they mean choosing an element according to the uniform distribution. On the other extreme, observe that for any $g \in G$, the function $\delta_g$ is a probability distribution for which $g$ has probability 1 of being chosen, and all other elements have no probability of being chosen whatsoever.

Notice that we can view a probability $P$ as an element of $L(G)$. A convenient fact is that the convolution of probabilities is again a probability. In fact, convolution has a very natural probabilistic interpretation. Let $P$ and $Q$ be probabilities on $G$. Suppose that one independently chooses $X$ at random according to $P$ and $Y$ at random according to $Q$. Let us compute the probability $XY = g$. If $Y = h$, then in order for $XY = g$ to occur, we must have $X = gh^{-1}$. The probability of these two events occurring simultaneously is $P(gh^{-1})Q(h)$ (by independence). Summing up over all possible choices of $h$ yields

$$\mathrm{Prob}[XY = g] = \sum_{h \in G} P(gh^{-1})Q(h) = P * Q(g).$$

Thus $P * Q$ is the distribution of the random variable $XY$ whenever $X$ and $Y$ are independent random variables with respective distributions $P$ and $Q$. The next proposition verifies formally that $P * Q$ is a probability distribution.

**Proposition 11.1.3.** *Let $P$ and $Q$ be probabilities on $G$. Then $P * Q$ is a probability on $G$ with support* $\mathrm{supp}(P * Q) = \mathrm{supp}(P) \cdot \mathrm{supp}(Q)$.

*Proof.* Trivially,

$$0 \leq \sum_{h \in G} P(gh^{-1})Q(h) \leq \sum_{h \in G} Q(h) = 1$$

and so $P * Q(g) \in [0, 1]$. Next, we compute

$$\begin{aligned}
\sum_{g \in G} P * Q(g) &= \sum_{g \in G} \sum_{h \in G} P(gh^{-1})Q(h) \\
&= \sum_{h \in G} Q(h) \sum_{g \in G} P(gh^{-1}) \\
&= \sum_{h \in G} Q(h) \\
&= 1
\end{aligned}$$

where the third equality uses that $gh^{-1}$ runs through each element of $G$ exactly once as $g$ does if $h$ is fixed. It follows that $P * Q$ is a probability distribution on $G$.

Notice that $P*Q(g) \neq 0$ if and only if there exists $h \in G$ such that $P(gh^{-1}) \neq 0$ and $Q(h) \neq 0$. Setting $x = gh^{-1}$ and $y = h$, it follows that $P * Q(g) \neq 0$ if and only if there exists $x \in \text{supp}(P)$ and $y \in \text{supp}(Q)$ such that $xy = g$. Thus, $\text{supp}(P * Q) = \text{supp}(P) \cdot \text{supp}(Q)$, as required. □

In order to measure how far a distribution is from being uniform, we need to define some notion of distance between probabilities. First, we introduce the $L^1$-norm on $L(G)$.

**Definition 11.1.4 ($L^1$-norm).** The $L^1$*-norm* on $L(G)$ is defined by

$$\|f\|_1 = \sum_{g \in G} |f(g)|$$

for $f\colon G \longrightarrow \mathbb{C}$.

For example, if $P$ is a probability on $G$, then $\|P\|_1 = 1$. The $L^1$-norm enjoys the following properties.

**Proposition 11.1.5.** *Let $a, b \in L(G)$. Then:*

1. $\|a\|_1 = 0$ *if and only if* $a = 0$;
2. $\|ca\|_1 = |c| \cdot \|a\|_1$ *for* $c \in \mathbb{C}$;
3. $\|a + b\|_1 \leq \|a\|_1 + \|b\|_1$ *(the triangle inequality);*
4. $\|a * b\|_1 \leq \|a\|_1 \cdot \|b\|_1$.

*Proof.* The first three items are immediate from the corresponding properties of the modulus of a complex number. We therefore just prove the final item, which is a straightforward computation:

$$\begin{aligned}
\|a * b\|_1 &= \sum_{g \in G} |a * b(g)| \\
&= \sum_{g \in G} \left| \sum_{h \in G} a(gh^{-1})b(h) \right| \\
&\leq \sum_{g \in G} \sum_{h \in G} |a(gh^{-1})||b(h)| \\
&= \sum_{h \in G} |b(h)| \sum_{g \in G} |a(gh^{-1})| \\
&= \|a\|_1 \cdot \|b\|_1
\end{aligned}$$

where the last equality uses that as $g$ varies through $G$, so does $gh^{-1}$ for fixed $h$. □

Probabilists use a variation of the $L^1$-norm to define the distance between probabilities. Let us first give a probabilistic definition as motivation.

**Definition 11.1.6 (Total variation).** The *total variation distance* between probabilities $P$ and $Q$ on a group $G$ is defined by

$$\|P-Q\|_{TV} = \max_{A\subseteq G} |P(A) - Q(A)|. \tag{11.2}$$

In plain language, two probabilities are close with respect to total variation distance if they differ by little on every subset of $G$. The total variation distance is closely related to the $L^1$-norm. To establish this relationship, we need the following lemma describing the set $A$ where the maximum in (11.2) is attained.

**Lemma 11.1.7.** *Let $P$ and $Q$ be probabilities on $G$. Let*

$$B = \{g \in G \mid P(g) \geq Q(g)\}$$
$$C = \{g \in G \mid Q(g) \geq P(g)\}.$$

*Then* $\|P-Q\|_{TV} = P(B) - Q(B) = Q(C) - P(C)$.

*Proof.* If $P = Q$, there is nothing to prove as all three quantities in question are zero. So we may assume that $P \neq Q$. Note that there must then be an element $g \in G$ such that $P(g) > Q(g)$. Indeed, if $P(g) \leq Q(g)$ for all $g \in G$, then $Q - P$ is non-negative on $G$ and

$$\sum_{g\in G}(Q(g) - P(g)) = \sum_{g\in G} Q(g) - \sum_{g\in G} P(g) = 1 - 1 = 0,$$

whence $P = Q$.

By definition, $\|P-Q\|_{TV} \geq |P(B) - Q(B)| = P(B) - Q(B)$. Let $A$ be a set such that $\|P-Q\|_{TV} = |P(A) - Q(A)|$ and let $A^c = G \setminus A$ be the complement of $A$. Then

$$|P(A^c) - Q(A^c)| = |1 - P(A) - (1 - Q(A))| = |Q(A) - P(A)| = \|P-Q\|_{TV}.$$

Let $g \in G$ with $P(g) > Q(g)$. Replacing $A$ by its complement if necessary, we may assume that $g \in A$. First we observe that $|P(A) - Q(A)| = P(A) - Q(A)$. For if, $|P(A) - Q(A)| = Q(A) - P(A)$, then we would have

$$Q(A \setminus \{g\}) - P(A \setminus \{g\}) > Q(A) - P(A) = \|P-Q\|_{TV},$$

a contradiction. Similarly, it follows that if $h \in A$ with $Q(h) > P(h)$, then $P(A \setminus \{h\}) - Q(A \setminus \{h\}) > P(A) - Q(A)$, again a contradiction. Thus $A \subseteq B$. But clearly $P(B) - Q(B) \geq P(A) - Q(A)$ for any subset $A \subseteq B$. Thus $\|P-Q\|_{TV} = P(B) - Q(B)$. A dual argument shows that $\|P-Q\|_{TV} = Q(C) - P(C)$. □

**Proposition 11.1.8.** *The equality*

$$\|P - Q\|_{TV} = \frac{1}{2}\|P - Q\|_1$$

*holds for all probabilities* $P, Q$ *on the finite group* $G$.

*Proof.* Let $B, C$ be as in Lemma 11.1.7. Then the lemma yields

$$\begin{aligned}
\|P - Q\|_{TV} &= \frac{1}{2}\left[P(B) - Q(B) + Q(C) - P(C)\right] \\
&= \frac{1}{2}\left[\sum_{\{g|P(g)\geq Q(g)\}} (P(g) - Q(g)) + \sum_{\{g|Q(g)\geq P(g)\}} (Q(g) - P(g))\right] \\
&= \frac{1}{2}\sum_{g\in G} |P(g) - Q(g)| \\
&= \frac{1}{2}\|P - Q\|_1
\end{aligned}$$

as required. □

Consequently, the total variation distance enjoys all the usual properties of a distance, such as the triangle inequality. As is usual in the theory of metric spaces, we say that a sequence $(P_n)$ of probabilities on $G$ *converges* to a probability $P$ if, for all $\varepsilon > 0$, there exists $N > 0$ such that $\|P_n - P\|_{TV} < \varepsilon$ whenever $n \geq N$.

## 11.2 Random Walks on Finite Groups

If $P$ is a probability on a group $G$, we write $P^{*k}$ for the $k$th convolution power of $P$ rather than $P^k$ in order to avoid confusion with the pointwise product of functions.

**Definition 11.2.1 (Random walk).** Let $P$ be a probability on a finite group $G$. Then the *random walk* on $G$ driven by $P$ is the sequence of probability distributions $(P^{*k})_{k=0}^{\infty}$.

The way to think about this is as follows. The random walker starts at the identity and chooses an element $X_1$ of $G$ according to $P$ and moves to $X_1$. Then he chooses an element $X_2$ according to $P$ and moves to $X_2X_1$, etc. Formally speaking, one considers a sequence of independent and identically distributed random variables $X_1, X_2, \ldots$ with common distribution $P$. Let $Y_0$ be the random variable with distribution $\delta_1$, that is, $Y_0 = 1$ with probability 1. Set $Y_k = X_kY_{k-1}$ for $k \geq 1$. The random variable $Y_k$ gives the position of the walker on the $k$th step of the random walk. Note that $Y_k$ is a random variable with distribution $P^{*k}$. So the random walk can be identified with either the sequence of random variables $Y_0, Y_1, \ldots$ or the sequence of probability distributions $\delta_1, P, P^{*2}, \ldots$. The sequence of random variables forms what is called in the probability literature a Markov chain [3].

*Example 11.2.2 (Simple random walk).* Let $G$ be a group and $S$ be a symmetric subset. Let $\Gamma$ be the Cayley graph of $G$ with respect to $S$. Then the *simple random walk* on $\Gamma$ is the random walk on $G$ driven by the probability $1/|S| \cdot \delta_S$. The walker starts at the identity of $G$. Suppose that at the $k$th step of the walk, the walker is at the vertex $g \in G$. Then an element $s \in S$ is chosen randomly (with all elements of $S$ equally probable) and the walker moves to the vertex $sg$. By construction, the vertices adjacent in $\Gamma$ to the vertex $g$ are precisely the elements of the form $sg$ with $s \in S$. So the group random walk formalism captures exactly the idea of the drunkard ambling randomly through the graph $\Gamma$!

One may also consider random walks on Cayley graphs where not all edges are equally probable.

*Example 11.2.3 (Discrete circle).* Let $0 \leq p, q \leq 1$ with $p + q = 1$. Suppose that one has a particle moving around the vertices of a regular $n$-gon. The particle moves one step clockwise with probability $p$ and one step counter-clockwise with probability $q$. This is the random walk on $\mathbb{Z}/n\mathbb{Z}$ driven by the probability $p\delta_{[1]} + q\delta_{[-1]}$.

Our next example is the Ehrenfest's urn, which is a model of diffusion.

*Example 11.2.4 (Ehrenfest's urn).* Suppose that one has two urns $A$ and $B$ containing a total of $n$ balls between them. At the beginning all the balls are in urn $A$. At each step in time, one of the $n$ balls is chosen at random (with all balls equally probable) and moved to the other urn. We can encode the configuration space by elements of $(\mathbb{Z}/2\mathbb{Z})^n$ as follows. If $v = (c_1, \ldots, c_n) \in (\mathbb{Z}/2\mathbb{Z})^n$, then the corresponding configuration has ball $i$ in urn $A$ if $c_i = [0]$ and ball $i$ in urn $B$ if $c_i = [1]$. So the initial configuration is the identity $([0], \ldots, [0])$. Let $e_i$ be the vector with $[1]$ in the $i$th-coordinate and with $[0]$ in all other coordinates. Then the configuration corresponding to $e_i + v$ is obtained from the configuration corresponding to $v$ by switching which urn contains ball $i$. Thus the stochastic process of switching the balls between urns corresponds to the random walk on $(\mathbb{Z}/2\mathbb{Z})^n$ driven by the probability

$$P = \frac{1}{n}(\delta_{e_1} + \cdots + \delta_{e_n}).$$

Equivalently, the Ehrenfest's urn walk can be considered as a simple random walk on the Cayley graph of $(\mathbb{Z}/2\mathbb{Z})^n$ with respect to the symmetric set $\{e_1, \ldots, e_n\}$. The case $n = 3$ appears in Fig. 11.1.

A random walk on a group $G$ can be viewed as a way to randomly generate an element of $G$. Normally, one wants the result to be unbiased, meaning that all elements of $G$ are equally likely. In other words, we would like to find a $k$ such that $\|P^{*k} - U\|_{TV}$ is very small. Preferably, we want $k$ not to be too large. As a first step, we would at least like $P^{*k}$ to converge to $U$. This is not always possible. Consider, for example, the simple random walk on $\mathbb{Z}/n\mathbb{Z}$ with respect to $S = \{\pm[1]\}$ in the case $n$ is even. Then clearly, at every even step of the walk, the walker is at

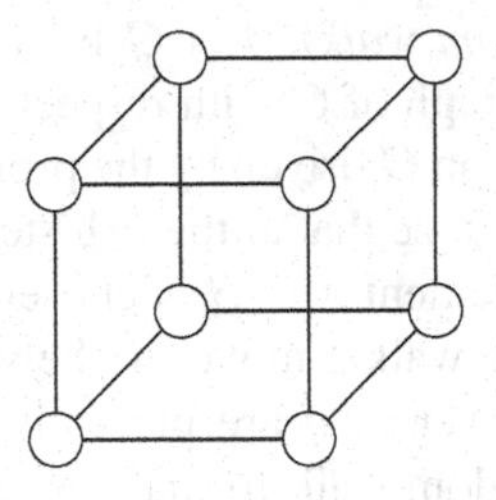

**Fig. 11.1** Cayley graph of $(\mathbb{Z}/2\mathbb{Z})^3$.

an element of the form $[m]$ with $m$ even and at every odd step the walker is at an element of the form $[m]$ with $m$ odd. Thus $P^{*k}$ never gets all that close to $U$ with respect to total variation. To rectify this periodicity issue, one often considers lazy random walks.

*Example 11.2.5 (Lazy random walk).* Let $S$ be a symmetric subset of the group $G$ and let $\Gamma$ be the corresponding Cayley graph. Suppose now that our drunkard is lazy, and so half the time he stays put where he is and the other half of the time he chooses to move along an edge to an adjacent vertex. This is modeled by the random walk on $G$ driven by the probability

$$P = \frac{1}{2}\delta_1 + \frac{1}{2|S|}\delta_S. \tag{11.3}$$

One calls this the *lazy random walk* on $\Gamma$. A common variant of the lazy random walk uses the probability

$$Q = \frac{1}{|S|+1}\delta_{S\cup\{1\}}.$$

One can think of this as corresponding to adding a loop edge at each vertex of the Cayley graph $\Gamma$.

In the next section we give a number of further examples coming from card shuffling.

We end this section by stating without proof a convergence theorem for random walks on finite groups.

**Definition 11.2.6 (Ergodic).** A random walk on a group $G$ driven by a probability $P$ is said to be *ergodic* if there exists an integer $N > 0$ such that $P^{*N}(g) > 0$ for all $g \in G$, i.e., $\mathrm{supp}(P^{*N}) = G$.

For instance, if $S$ is a symmetric generating set for $G$, then the lazy random walk on the Cayley graph of $G$ with respect to $S$ is an ergodic random walk, as the following proposition shows. It is precisely for this reason that we have introduced the "laziness."

**Proposition 11.2.7.** *Let $P$ be a probability on a finite group $G$ and suppose that:*

*1.* $P(1) > 0$*;*
*2.* $\mathrm{supp}(P)$ *generates the group $G$.*

*Then the random walk driven by $P$ is ergodic.*

*Proof.* Let $S = \mathrm{supp}(P)$. Note that $S^k \subseteq S^{k+1}$ for all $k \geq 0$ because $1 \in S$. Thus, as $S$ is a generating set for $G$, there exists $N > 0$ such that $S^N = G$. By Proposition 11.1.3, the support of $P^{*N}$ is $G$, that is, $P^{*N}(g) > 0$ for all $g \in G$. This yields the ergodicity of the random walk. □

The simple random walk on $\mathbb{Z}/n\mathbb{Z}$ with respect to $S = \{\pm[1]\}$ is never ergodic when $n$ is even because $P^{*k}$ has support on the odd elements when $k$ is odd and on the even elements when $k$ is even. (The reader should check that it is ergodic when $k$ is odd!) The Ehrenfest's urn random walk is also not ergodic since on even steps of the walk, the probability is supported on vectors with an even number of coordinates equal to $[1]$, whereas on odd steps of the walk there will always be an odd number of coordinates equal to $[1]$. So often one considers instead a lazy random walk on the Cayley graph of $(\mathbb{Z}/2\mathbb{Z})^n$ with respect to $\{e_1, \ldots, e_n\}$.

**Theorem 11.2.8.** *Let $(P^{*k})_{k=0}^{\infty}$ be an ergodic random walk on a finite group $G$ driven by the probability $P$. Then the sequence $(P^{*k})$ converges to the uniform distribution $U$.*

This theorem means intuitively that an ergodic random walk on a finite group $G$ can be used to randomly generate elements of $G$. We shall prove Theorem 11.2.8 for abelian groups in Section 11.4. The reader is referred to [3, 7] for the general case.

## 11.3 Card Shuffling

In this section, we give a short introduction to the mathematics of card shuffling viewed as random walks on the symmetric group. For a more detailed account from the probabilistic viewpoint, the reader is referred to [18]. Our presentation follows mostly [7, Chapter 4].

Suppose that we have a deck of $n$ cards. Then performing a shuffle to the cards means in simplest terms to reorder them in some way. This corresponds to the action of an element of the symmetric group $S_n$. We view the symmetric group as acting on the positions of the cards, not the names of the cards. For example, the permutation $(3\ 2\ 1)$ corresponds to putting the top card into the third position. This moves the second card to the top and the third card to the second position. The remaining cards are left alone. This leads us to our first example of a card shuffling method as a random walk.

*Example 11.3.1 (Top-to-random).* The *top-to-random shuffle* takes the top card from the deck and places it into one of the $n$ positions of the deck at random (with all positions equally probable). Placing the top card into the top position corresponds to the identity map, of course. For $i \geq 2$, putting the top card in the $i$th position shifts the previous positions up one and so corresponds to the cycle $(i\; i-1 \cdots 1)$. Thus the top-to-random shuffle can be modeled by the random walk on $S_n$ driven by the probability

$$P = \frac{1}{n}\delta_{Id} + \sum_{i=2}^{n} \frac{1}{n}\delta_{(i\; i-1 \cdots 1)}.$$

The permutations $(2\,1)$ and $(n\; n-1 \cdots 1)$ generate $S_n$, hence this walk is ergodic by Proposition 11.2.7. This means that after enough top-to-random shuffles, the deck will eventually be mixed.

The next shuffling method we consider is that of random transpositions. Convergence times for this method were first obtained by Diaconis and Shahshahani using the representation theory of the symmetric group [7,9]. The model works as follows.

*Example 11.3.2 (Random transpositions).* The dealer randomly picks with each of his hands a card from the deck (possibly both hands pick the same card). He then swaps the two cards (and so if he picked the same card with each hand, then he does nothing). Given two positions $i \neq j$, there are two ways the dealer can pick this pair (either the left hand picks $i$ and the right hand picks $j$, or vice versa) and so the probability of performing the transposition $(i\; j)$ is $2/n^2$. The probability that the dealer picks position $i$ with both hands is $1/n^2$. However, the resulting permutation of the positions is the identity for all $i$ and so we perform the identity permutation with probability $1/n$. Thus the *random transpositions shuffle* is the random walk on $S_n$ driven by the probability $Q$ defined by

$$Q(\sigma) = \begin{cases} 1/n & \sigma = Id \\ 2/n^2 & \sigma \text{ is a transposition} \\ 0 & \text{else.} \end{cases}$$

Since the transpositions generate $S_n$, Proposition 11.2.7 implies that this is an ergodic random walk and so again this shuffle will randomize the deck.

### *11.3.1 The Riffle Shuffle*

Nobody really shuffles a deck by randomly swapping cards. Perhaps some young children shuffle via the top-to-random method, but not serious card players. The most commonly used shuffle in practice is the *riffle shuffle*, also known as the *dovetail shuffle*. In this shuffle, the dealer cuts the pack somewhere near the middle and then places the top half of the pack in his right hand and the bottom half in his left hand. He then drops cards from each packet, interleaving the two packets. In a perfect shuffle, the dealer would drop one card from each packet in alternation, but in reality several cards from the same packet are often dropped at a time.

A mathematical model of riffle shuffling was proposed by Gilbert and Shannon, and independently by Reeds, and is known as the *Gilbert–Shannon–Reeds shuffle*. The model works as follows. Suppose that the deck has $n$ cards. Then one flips a fair coin $n$ times; if the number of heads is $k$, then the dealer places the top $k$ cards in his right hand and the bottom $n-k$ cards in his left hand. In technical language, this means that the position of the cut is assumed to be a binomial random variable. So if $X$ is the random variable counting the number of cards in the top half of the deck after the cut, then

$$\text{Prob}[X = k] = \binom{n}{k}\frac{1}{2^n}.$$

Notice that the closer $k$ is to $n/2$, the more likely $X = k$ is to happen. The reader can easily convince himself of this by drawing Pascal's triangle and observing that the binomial coefficients are largest in the middle of each row. The model thus reflects the fact that we tend to cut the deck close to the middle.

Next the dealer drops cards from the left and right hand until both hands are empty with the probability of dropping a card from a given hand being proportional to the number of cards in that hand. If there are $a$ cards in the right hand and $b$ cards in the left hand at a given moment, then the probability of dropping a card from the right hand is $a/(a+b)$ and the probability of dropping a card from the left hand is $b/(a+b)$. Notice that this model allows the possibility that all the cards are dropped from the left hand first, in which case the ordering of the deck remains as it was before. Also, if $k = 0$ or $k = n$, then we did not really cut the deck and so the result is that the cards remain in their original order. One might argue that this does not happen in reality, but in any event the probability of this occurring is still quite small in this model.

See [7, Chap. 4] for some discussion on the accuracy of the Gilbert–Shannon–Reeds model. Here, we want to describe the Gilbert–Shannon–Reeds shuffle as a random walk on $S_n$. The key to understanding riffle shuffles is the notion of a rising sequence.

**Definition 11.3.3 (Rising sequence).** A *rising sequence*[1] in a permutation $\sigma$ of $\{1,\ldots,n\}$ is a maximal increasing, consecutive subsequence in the image sequence $\sigma(1),\sigma(2),\ldots,\sigma(n)$.

This notion is best illustrated with an example.

*Example 11.3.4.* Let $\sigma = (1\ 2\ 3)(7\ 4\ 8)(5\ 6) \in S_8$. Then the sequence of images is $2,3,1,8,6,5,4,7$. The rising sequences of $\sigma$ are: $2,3$; $1,8$; $6$; $5$; and $4,7$. Thus $\sigma$ has five rising sequences.

[1] Some authors use a different convention to relate permutations to shuffles and hence use a different definition of rising sequence. We follow [7].

Suppose now that one performs a riffle shuffle where the top half of the deck has $k$ cards. Then the cards in positions $1, \ldots, k$ are interleaved with the cards in positions $k+1, \ldots, n$, with the relative ordering of the cards preserved. Thus the resulting permutation $\sigma$ of the positions will be increasing on $1, \ldots, k$ and on $k+1, \ldots, n$, and hence have exactly two rising sequences: $\sigma(1), \ldots, \sigma(k)$ and $\sigma(k+1), \ldots, \sigma(n)$ (except for in the case that the bottom half of the deck is dropped in its entirety before cards from the top half are dropped, in which case we obtain the identity permutation, which has 1 rising sequence). Let us illustrate this on an example.

*Example 11.3.5.* Suppose our deck has ten cards in the order, from top to bottom, $\mathsf{A}, 2, 3, 4, 5, 6, 7, 8, 9, 10$ and we perform a riffle shuffle with four cards – $\mathsf{A}, 2, 3, 4$ – in the top packet and six cards – $5, 6, 7, 8, 9, 10$ – in the bottom packet. Suppose that the sequence of drops is $T, B, B, T, B, T, T, B, B, B$ where $T$ stands for "top" and $B$ for "bottom." Then the four drops first, putting it at the bottom of the deck, the ten drops second putting it ninth in the deck and so on. So in the shuffled deck the order of cards will be (again, from top to bottom) $5, 6, 7, \mathsf{A}, 2, 8, 3, 9, 10, 4$, which corresponds to the permutation $\sigma$ of the positions with image sequence $4, 5, 7, 10, 1, 2, 3, 6, 8, 9$ (as the card in the first position, A, is now in the fourth position, etc.). The rising sequences of $\sigma$ are $4, 5, 7, 10$ and $1, 2, 3, 6, 8, 9$.

Therefore, if $P$ is the probability distribution on $S_n$ corresponding to a Gilbert–Shannon–Reeds shuffle, then $P(\sigma) = 0$ unless $\sigma$ has at most two rising sequences. It remains to compute the probability that a given permutation with two rising sequences is obtained, as well as the probability of obtaining the identity.

Let us condition on the number of cards $k$ in the top half of the deck after the cut. There are $n$ positions to place these $k$ cards and once these positions are chosen, the ordering of the cards is known. Thus there are $\binom{n}{k}$ possible permutations obtainable given $k$. We claim that each of them is equally probable.

Suppose that one fixes such a permutation $\sigma$. What is its probability? Remember our model: the probability of each drop being from the top packet, let us call this event $T$, is $a/(a+b)$ and from the bottom packet, let us call this event $B$, is $b/(a+b)$ where $a$ is the number of cards in the top packet and $b$ is the number of cards in the bottom packet. So when the cut is $k$ cards in the top packet and $n-k$ cards in the bottom packet, then the probability that the sequence of drops starts with $T$ is $k/n$ and the probability that the sequence starts with a $B$ is $(n-k)/n$. Regardless of whether the first drop is a $T$ or a $B$, the denominator of the probability for the second drop will be reduced by 1 to $n-1$ and the numerator for the one chosen in the first position will be reduced by 1 and be either $k-1$ or $n-k-1$, according to whether the top or bottom card was dropped, respectively. This will continue for each succeeding position so the numbers $n, n-1, n-2, \ldots, 1$ will appear in the denominators and the numbers $k, k-1 \ldots, 1$ and $n-k, n-k-1, \ldots, 1$ will all appear in some one of the numerators. Since the probability of any sequence of drops is the product of these individual probabilities it will in every case be $k!(n-k)!/n!$ and so they are all equally probable.

To make things more concrete, let us analyze a small example. Suppose that there are five cards and $k = 2$. In this case $k!(n-k)!/n! = 2! \cdot 3!/5! = 1/10$. Now we will consider two sequences of drops. First, consider the sequence $T, B, T, B, B$. According to the model its probability is

$$\frac{2}{5} \cdot \frac{3}{4} \cdot \frac{1}{3} \cdot \frac{2}{2} \cdot \frac{1}{1} = 1/10.$$

As a second example, we see that the sequence $B, T, B, T, B$ has the probability

$$\frac{3}{5} \cdot \frac{2}{4} \cdot \frac{2}{3} \cdot \frac{1}{2} \cdot \frac{1}{1} = 1/10,$$

as well.

In summary, the probability of obtaining a given permutation $\sigma$, which is obtainable when the cut is in position $k$, is

$$\frac{k!(n-k)!}{n!} = \frac{1}{\binom{n}{k}}.$$

Exactly one of these permutations will be the identity. The others will have two rising sequences: one of length $k$, followed by another of length $n-k$. Since the probability that there will be $k$ cards in the top half is $\binom{n}{k} \cdot 1/2^n$, it follows that if $1 \leq k \leq n-1$ and $\sigma$ is a permutation with rising sequences $\sigma(1), \ldots, \sigma(k)$ and $\sigma(k+1), \ldots, \sigma(n)$, then the probability of obtaining $\sigma$ in a single riffle shuffle is

$$\binom{n}{k} \frac{1}{2^n} \cdot \frac{1}{\binom{n}{k}} = \frac{1}{2^n}.$$

On the other hand, the same argument shows that, for any $k$, the probability of obtaining the identity permutation is also $1/2^n$. Letting $k$ range from 0 to $n$, we conclude that the probability of getting the identity permutation (that is, of not shuffling the deck at all) is $(n+1)/2^n$. Thus we have the following model of the Gilbert–Shannon–Reeds shuffle as a random walk on $S_n$.

**Proposition 11.3.6.** *The Gilbert–Shannon–Reeds shuffle corresponds to the random walk on $S_n$ driven by the probability distribution $P$ given by*

$$P(\sigma) = \begin{cases} (n+1)/2^n & \sigma = Id \\ 1/2^n & \sigma \text{ has exactly two rising sequences} \\ 0 & \text{else.} \end{cases}$$

Observe that the permutations $(1\ 2)$ and $(1\ 2 \cdots n)$ both have exactly two rising sequences. If we cut the pack at the top card and drop all the cards but one from the bottom half first, then we obtain the permutation $(1\ 2)$. If we cut the deck at the last card and drop the entire top half first, then the resulting permutation of the positions is $(1\ 2 \cdots n)$. Thus the random walk associated to the Gilbert–Shannon–Reeds model is ergodic by Proposition 11.2.7 and so indeed repeated riffle shuffling will randomize the deck!

## 11.4 The Spectrum and the Upper Bound Lemma

Fix a finite group $G$ and let $P$ be a probability on $G$. Analyzing the random walk on $G$ driven by $P$ amounts to understanding the convolution powers of $P$. Associated to $P$ is the convolution operator $M\colon L(G) \longrightarrow L(G)$ given by $M(a) = P * a$. In particular, $M^k(\delta_1) = P^{*k}$ and so studying the powers of $M$ is tantamount to studying the random walk. Of course, a key ingredient in the qualitative study of the powers of a linear operator is its spectrum.

**Definition 11.4.1 (Spectrum).** The *spectrum* of the random walk on a finite group $G$ driven by the probability $P$ is the set of eigenvalues (with multiplicities) of the linear operator $M\colon L(G) \longrightarrow L(G)$ given by $M(a) = P * a$. The spectrum is denoted $\operatorname{spec}(P)$.

It can be shown (see Exercise 11.1) that $P * U = U$ and so $U$ is an eigenvector of $M$ with eigenvalue 1. We call the eigenvalue 1 trivial. One can show that all eigenvalues in $\operatorname{spec}(P)$ have modulus at most 1 (see Exercise 11.6).

The spectrum is quite easy to understand in the case of an abelian group via Fourier analysis.

**Theorem 11.4.2 (Diaconis [7]).** *Let $G$ be a finite abelian group and let $P$ be a probability on $G$. Then $\operatorname{spec}(P) = \{\widehat{P}(\chi) \mid \chi \in \widehat{G}\}$ where the multiplicity of $\lambda \in \operatorname{spec}(P)$ is the number of characters $\chi$ for which $\widehat{P}(\chi) = \lambda$. An orthonormal basis for the eigenspace of $\lambda$ is the set of all characters $\chi$ with $\widehat{P}(\chi) = \lambda$.*

*Proof.* This is a special case of Lemma 5.4.9. □

Notice that $U = 1/|G| \cdot \chi_1$ where $\chi_1$ is the character of the trivial representation. Moreover,

$$\widehat{P}(\chi_1) = |G|\langle P, \chi_1\rangle = \sum_{g\in G} P(g) = 1 \tag{11.4}$$

yielding the eigenvalue 1. If $\chi$ is any irreducible character, then

$$|\widehat{P}(\chi)| = |G| \cdot |\langle P, \chi\rangle| = \left|\sum_{g\in G} P(g)\overline{\chi(g)}\right| \leq \sum_{g\in G} P(g)|\overline{\chi(g)}| = 1 \tag{11.5}$$

where the last equality uses that $\chi(g)$ has modulus 1. Thus

$$\operatorname{spec}(P) \subseteq \{z \in \mathbb{C} \mid |z| \leq 1\}.$$

Let us compute the spectrum in some examples.

*Example 11.4.3.* Consider the lazy random walk on $\mathbb{Z}/n\mathbb{Z}$ driven by the probability measure

$$P = \frac{1}{2}\delta_{[0]} + \frac{1}{4}\delta_{[1]} + \frac{1}{4}\delta_{[-1]}.$$

As usual, define $\chi_k([m]) = e^{2\pi ikm/n}$. Then the eigenvalue $\widehat{P}(\chi_k)$ is given by

$$\widehat{P}(\chi_k) = \frac{1}{2} + \frac{1}{4}\left[e^{-2\pi ik/n} + e^{2\pi ik/n}\right] = \frac{1}{2} + \frac{1}{2}\cos 2\pi k/n.$$

Our next example computes the spectrum of the Ehrenfest's urn walk.

*Example 11.4.4.* The irreducible characters of $G = (\mathbb{Z}/2\mathbb{Z})^n$ admit the following description, as can easily be verified using the results of Section 4.5. If $v = (c_1, \ldots, c_n) \in G$, define $\alpha(v) = \{i \mid c_i = [1]\}$. Given $Y \subseteq \{1, \ldots, n\}$, define $\chi_Y : G \longrightarrow \mathbb{C}$ by

$$\chi_Y(v) = (-1)^{|\alpha(v) \cap Y|}.$$

Then $\widehat{G} = \{\chi_Y \mid Y \subseteq \{1, \ldots, n\}\}$.

The probability driving the Ehrenfest's urn walk is $P = 1/n(\delta_{e_1} + \cdots + \delta_{e_n})$ and so the eigenvalue corresponding to $\chi_Y$ is given by $\widehat{P}(\chi_Y) = n - 2|Y|/n$. Indeed,

$$\chi_Y(e_i) = \begin{cases} -1 & i \in Y \\ 1 & \text{else.} \end{cases}$$

Thus $|Y|$ elements of $\{e_1, \ldots, e_n\}$ contribute a value of $-1/n$ to $\widehat{P}(\chi_Y)$, whereas the remaining $n - |Y|$ elements contribute $1/n$. Thus the spectrum of the Ehrenfest's urn walk consists of the numbers $\lambda_j = 1 - 2j/n$ with $0 \leq j \leq n$. The multiplicity of $\lambda_j$ is $\binom{n}{j}$ because this is the number of subsets $Y$ with $j$ elements.

Our next goal is to give an upper bound, due to Diaconis and Shahshahani [3,7,9], on the distance from a probability on an abelian group to the uniform distribution in terms of the Fourier transform of the probability (or equivalently, the spectrum of the random walk). This is particularly convenient since the Fourier transform turns convolution powers to pointwise products. First we need a lemma relating the $L^1$-norm to the norm coming from the inner product: $\|f\| = \sqrt{\langle f, f \rangle}$.

**Lemma 11.4.5.** *Let $G$ be a finite group and $f \in L(G)$. Then $\|f\|_1 \leq |G| \cdot \|f\|$.*

*Proof.* Let $\chi_1$ be the trivial character: so $\chi_1(g) = 1$ for all $g \in G$. Write $|f|$ for the function given by $|f|(g) = |f(g)|$ for $g \in G$. Then

$$\|f\|_1 = |G| \cdot \langle |f|, \chi_1 \rangle \leq |G| \cdot \|f\| \cdot \|\chi_1\| = |G| \cdot \|f\|$$

where the inequality is the Cauchy–Schwarz inequality and we have used that $\|\chi_1\| = 1$. □

Next, we must relate the norm of the Fourier transform of a function to its original norm.

**Theorem 11.4.6 (Plancherel).** *Let $G$ be a finite abelian group and let $a, b \in L(G)$. Then*

$$\langle a, b\rangle = \frac{1}{|G|}\langle \widehat{a}, \widehat{b}\rangle.$$

*In particular, $\|a\|^2 = \|\widehat{a}\|^2/|G|$.*

*Proof.* By the Fourier inversion theorem, we have

$$a = \frac{1}{|G|}\sum_{\chi\in\widehat{G}} \widehat{a}(\chi)\chi$$

$$b = \frac{1}{|G|}\sum_{\chi\in\widehat{G}} \widehat{b}(\chi)\chi.$$

Using the orthonormality of the irreducible characters, we conclude

$$\langle a, b\rangle = \frac{1}{|G|^2}\sum_{\chi\in\widehat{G}} \widehat{a}(\chi)\overline{\widehat{b}(\chi)} = \frac{1}{|G|}\langle \widehat{a}, \widehat{b}\rangle$$

since $|\widehat{G}| = |G|$. □

We are now ready to prove the Diaconis–Shahshahani upper bound lemma from [9] (for the case of abelian groups).

**Lemma 11.4.7 (Upper bound lemma).** *Let $G$ be a finite abelian group and let $\widehat{G}^*$ denote the set of non-trivial irreducible characters of $G$. Let $Q$ be a probability on $G$. Then*

$$\|Q - U\|_{TV}^2 \leq \frac{1}{4}\sum_{\chi\in\widehat{G}^*} |\widehat{Q}(\chi)|^2.$$

*Proof.* Applying Proposition 11.1.8 and Lemma 11.4.5, we have

$$\|Q - U\|_{TV}^2 = \frac{1}{4}\|Q - U\|_1^2 \leq \frac{1}{4}|G|^2\|Q - U\|^2. \tag{11.6}$$

By the Plancherel formula, we have

$$|G|^2\|Q - U\|^2 = |G|\|\widehat{Q} - \widehat{U}\|^2 = |G|\left[\langle\widehat{Q},\widehat{Q}\rangle - 2\langle\widehat{Q},\widehat{U}\rangle + \langle\widehat{U},\widehat{U}\rangle\right]. \tag{11.7}$$

Now $\widehat{U} = \delta_{\chi_1}$, where $\chi_1$ is the trivial character, by the orthogonality relations because

$$\widehat{U}(\chi) = |G|\langle U, \chi\rangle = |G|\langle 1/|G|\cdot\chi_1, \chi\rangle = \langle\chi_1, \chi\rangle$$

for $\chi \in \widehat{G}$. Also, $\widehat{P}(\chi_1) = 1$ for any probability $P$ by (11.4). Thus $\langle \widehat{U}, \widehat{U} \rangle = 1/|G|$, $\langle \widehat{Q}, \widehat{U} \rangle = \widehat{Q}(\chi_1) = 1/|G|$ and

$$\langle \widehat{Q}, \widehat{Q} \rangle = 1/|G| + 1/|G| \sum_{\chi \in \widehat{G}^*} \widehat{Q}(\chi)\overline{\widehat{Q}(\chi)}.$$

Substituting into (11.7), we obtain

$$|G|^2 \|Q - U\|^2 = \sum_{\chi \in \widehat{G}^*} |\widehat{Q}(\chi)|^2$$

from which the result follows in light of (11.6). □

An important special case is when $Q = P^{*k}$ where $P$ is a probability on $G$. Using that the Fourier transform turns convolution into pointwise product, we obtain the following corollary.

**Corollary 11.4.8.** *Let $G$ be a finite abelian group and let $P$ be a probability on $G$. Then*

$$\|P^{*k} - U\|_{TV}^2 \le \frac{1}{4} \sum_{\chi \in \widehat{G}^*} |\widehat{P}(\chi)|^{2k}$$

*where $\widehat{G}^*$ denotes the set of non-trivial characters of $G$.*

*Remark 11.4.9.* Notice that the complex numbers $\widehat{P}(\chi)$ with $\chi \in \widehat{G}^*$ are the non-trivial elements of the spectrum of $P$ provided 1 is a simple eigenvalue, which is the case if, for instance, the random walk is ergodic.

The upper bound lemma can be used to obtain very tight bounds on the rate of convergence for a variety of random walks. Let us consider in some detail the lazy version of the Ehrenfest's urn, Example 11.2.4, driven by the probability

$$P = \frac{1}{n+1} \left[ \delta_{(0,\ldots,0)} + \delta_{e_1} + \cdots + \delta_{e_n} \right]. \tag{11.8}$$

**Theorem 11.4.10 (Diaconis [7]).** *Let $P$ be the probability distribution on $(\mathbb{Z}/2\mathbb{Z})^n$ from (11.8) and let $c > 0$ be a positive constant. Then, for $k \ge (n+1)(\log n + c)/4$, the inequality*

$$\|P^{*k} - U\|_{TV}^2 \le \frac{1}{2}(e^{e^{-c}} - 1)$$

*holds.*

*On the other hand, if $k \le (n+1)(\log n - c)/4$ where $0 < c < \log n$ and $n$ is sufficiently large, then*

$$\|P^{*k} - U\|_{TV} \ge 1 - 20e^{-c}$$

*holds.*

Observe that $\sqrt{(e^{e^{-c}}-1)/2}$ goes to zero extremely quickly as $c$ goes to infinity, whereas $1-20e^{-c}$ goes to 1 very quickly. For example, taking $c=10$ yields $\sqrt{(e^{e^{-10}}-1)/2}\approx 0.004765$ and $1-20e^{-10}\approx 0.999092$. Roughly speaking, Theorem 11.4.10 says that in $1/4\cdot(n+1)\log n$ steps, all possible configurations of the balls in the two urns are almost equally probable, but in any fewer steps, this is very far from the case. This phenomenon of changing very rapidly from non-uniform to uniform behavior is called by Diaconis the *cut-off phenomenon*.

We shall content ourselves with a proof of the upper bound in Theorem 11.4.10, as the lower bound requires probabilistic arguments beyond the scope of this text.

First we need two lemmas. Recall that if $x$ is a real number, then $\lfloor x\rfloor$ is the largest integer below $x$, that is, the integer you get by rounding down $x$. For example, $\lfloor 5/2\rfloor = 2$.

**Lemma 11.4.11.** *Suppose that* $1\leq j\leq\lfloor n+1/2\rfloor$. *Then* $\binom{n}{j-1}\leq\binom{n}{j}$.

*Proof.* The above inequality is equivalent to

$$\frac{n!}{(j-1)!(n-j+1)!}\leq\frac{n!}{j!(n-j)!},$$

which is in turn equivalent to the inequality $j\leq n-j+1$. But this occurs if and only if $j\leq(n+1)/2$, which is true by assumption. □

Our second lemma concerns the exponential function.

**Lemma 11.4.12.** *If* $0\leq x\leq 1$, *then* $(1-x)^{2k}\leq e^{-2kx}$ *for all* $k\geq 0$.

*Proof.* First we prove for $x\in\mathbb{R}$ that $1-x\leq e^{-x}$. Let $f(x)=e^{-x}-(1-x)=e^{-x}-1+x$. Then $f'(x)=-e^{-x}+1$. Thus $f'(x)<0$ when $x<0$, $f'(0)=0$ and $f'(x)>0$ when $x>0$. It follows that $f$ achieves its global minimum at $x=0$. But $f(0)=0$ and so $e^{-x}\geq 1-x$ for all $x\in\mathbb{R}$. As a consequence, for $0\leq x\leq 1$, we have $(1-x)^{2k}\leq e^{-2kx}$. □

We are now prepared to prove the upper bound in Theorem 11.4.10.

*Proof (of Theorem 11.4.10).* Let $Y\subseteq\{1,\ldots,n\}$ and let $\chi_Y$ be the corresponding character of $(\mathbb{Z}/2\mathbb{Z})^n$ as per Example 11.4.4. Then the same argument as in that example yields

$$\widehat{P}(\chi_Y)=\frac{n+1-2|Y|}{n+1}=1-\frac{2|Y|}{n+1}. \tag{11.9}$$

The trivial character corresponds to $Y=\emptyset$. As there are $\binom{n}{j}$ subsets $Y$ with $|Y|=j$, the corollary to the upper bound lemma (Corollary 11.4.8), in conjunction with (11.9), implies

$$\|P^{*k}-U\|_{TV}^2\leq\frac{1}{4}\sum_{j=1}^{n}\binom{n}{j}\left[1-\frac{2j}{n+1}\right]^{2k}. \tag{11.10}$$

Suppose now that $j+\ell=n+1$. Then we have that

$$1-\frac{2j}{n+1}=1-\frac{2n+2-2\ell}{n+1}=\frac{2\ell}{n+1}-1 \tag{11.11}$$

$$\binom{n}{\ell}=\binom{n}{n-(j-1)}=\binom{n}{j-1}. \tag{11.12}$$

Substituting (11.11) and (11.12) into (11.10) and applying Lemma 11.4.11 yields

$$\|P^{*k}-U\|_{TV}^2\leq\frac{1}{2}\sum_{j=1}^{\lfloor\frac{n+1}{2}\rfloor}\binom{n}{j}\left[1-\frac{2j}{n+1}\right]^{2k}. \tag{11.13}$$

Trivially,

$$\binom{n}{j}=\frac{n!}{j!(n-j)!}=\frac{n(n-1)\cdots(n-j+1)}{j!}\leq\frac{n^j}{j!}$$

and so in conjunction with Lemma 11.4.12 (with $x=2j/(n+1)$), we obtain from (11.13) the inequality

$$\|P^{*k}-U\|_{TV}^2\leq\frac{1}{2}\sum_{j=1}^{\lfloor\frac{n+1}{2}\rfloor}\frac{n^j}{j!}e^{-\frac{4kj}{n+1}}. \tag{11.14}$$

Assume now that $k\geq(n+1)(\log n+c)/4$. Then

$$e^{-\frac{4kj}{n+1}}\leq e^{-j\log n-jc}=e^{-jc}/n^j.$$

Applying this to (11.14) yields

$$\begin{aligned}\|P^{*k}-U\|_{TV}^2&\leq\frac{1}{2}\sum_{j=1}^{\lfloor\frac{n+1}{2}\rfloor}\frac{1}{j!}e^{-jc}\\&\leq\frac{1}{2}\sum_{j=1}^{\infty}\frac{1}{j!}(e^{-c})^j\\&=\frac{1}{2}\left(e^{e^{-c}}-1\right)\end{aligned}$$

as required. □

Simple random walk on the discrete circle is studied in detail in [7, Chap. 3] and [3]. The corresponding result to Theorem 11.4.10, whose proof can be found in these sources, is the following.

**Theorem 11.4.13 (Diaconis [7]).** *For $n$ odd, if $P$ is the probability on $\mathbb{Z}/n\mathbb{Z}$ given by $P = 1/2 \cdot (\delta_{[1]} + \delta_{[-1]})$, then*

$$\|P^{*k} - U\|_{TV} \leq e^{-\frac{\pi^2 k}{2n^2}}$$

*for $k \geq n^2$.*

*For $n \geq 6$ and all $k \geq 0$, the inequality*

$$\|P^{*k} - U\|_{TV} \geq \frac{1}{2} e^{-\frac{\pi^2 k}{2n^2} - \frac{\pi^4 k}{2n^4}}$$

*holds.*

Theorem 11.4.13 does not exhibit the same cut-off phenomenon we saw earlier for the Ehrenfest's urn walk. We remind the reader that, for $n$ even, the simple random walk on the discrete circle does not converge to the uniform distribution.

Diaconis and Shahshahani [7, 9] actually proved an upper bound lemma for arbitrary finite groups. Let $G$ be a finite group and let $\varphi^{(2)}, \ldots, \varphi^{(s)}$ be a complete set of unitary representatives of the non-trivial irreducible representations of $G$. Let $d_i$ be the degree of $\varphi^{(i)}$. Then one can show using the analog of the Plancherel formula for non-abelian groups (cf. Exercise 5.7) that the bound

$$\|Q - U\|_{TV}^2 \leq \frac{1}{4} \sum_{i=2}^{s} d_i \operatorname{Tr}\left[\widehat{Q}(\varphi^{(i)})\widehat{Q}(\varphi^{(i)})^*\right] \tag{11.15}$$

holds for any probability $Q$ on $G$. Notice that if $Q$ is a class function, then $\widehat{Q}(\varphi^{(i)})$ is a scalar matrix by Schur's lemma. This makes (11.15) easier to apply in this context. In particular, this applies to card shuffling by random transpositions.

*Example 11.4.14 (Card shuffling).* We indicate here, without proof, the rates of convergence for the various card shuffling walks discussed earlier. The reader is referred to [3, 7, 18] for details.

In a deck with $n$ cards, shuffling by random transpositions has a sharp cut-off at $\frac{1}{2}n \log n$ shuffles. The top-to-random shuffle has a sharp cut-off at $n \log n$ shuffles. The cut-off for the riffle shuffle is $\frac{3}{2} \log_2 n$ (where the log here is base 2.). For a deck of 52 cards, this yields approximately 44.62 shuffles for random transpositions, 89.23 shuffles for the top-to-random shuffle and 8.55 shuffles for riffle shuffling. However, Bayer and Diaconis argue [1] that seven is really enough for riffle shuffling.

*Remark 11.4.15 (Gelfand pairs).* There is an analog of the upper bound lemma in the context of Gelfand pairs [3, Corollary 4.9.2] that is useful for studying random walks on coset spaces of groups.

We end this section by applying the upper bound lemma to prove that ergodic random walks on finite abelian groups converge to the uniform distribution.

**Theorem 11.4.16.** *Let $P$ be a probability on a finite abelian group $G$ and suppose that the random walk driven by $P$ is ergodic. Then the sequence $(P^{*n})$ converges to the uniform distribution.*

*Proof.* In light of Corollary 11.4.8, it suffices to show that $|\widehat{P}(\chi)| < 1$ for all non-trivial characters $\chi$ of $G$. This occurs if and only if $|\widehat{P}(\chi)|^m < 1$ for some $m > 0$ and so without loss of generality, we may assume that $P(g) > 0$ for all $g \in G$ (using the definition of ergodicity and that the Fourier transform turns convolution into the pointwise product). Recall from the proof of Lemma 6.3.1 that, for complex numbers $\lambda_1, \ldots, \lambda_r$, one has $|\lambda_1 + \cdots + \lambda_r| = |\lambda_1| + \cdots + |\lambda_r|$ if and only if the $\lambda_i$ are non-negative multiples of some complex number. In light of this, consideration of (11.5) shows that $|\widehat{P}(\chi)| \leq 1$ and that if equality holds, then the numbers $P(g)\overline{\chi(g)}$ with $g \in G$ are all non-negative multiples of a common complex number. But $P(1) > 0$ and $\overline{\chi(1)} = 1$. On the other hand, since $\chi$ is non-trivial, also $\overline{\chi(g)} \neq 1$ for some $g \in G$. Because $P(g) > 0$ and $\overline{\chi(g)}$ is a non-trivial $|G|$th-root of unity, it follows that $P(1)\overline{\chi(1)} = P(1)$ and $P(g)\overline{\chi(g)}$ are not non-negative multiples of a common complex number and so the inequality in (11.5) is strict. Thus, $|\widehat{P}(\chi)| < 1$, as was required. This completes the proof. □

## Exercises

**Exercise 11.1.** Show that if $P$ is any probability on a finite group $G$, then $P * U = U = U * P$. Deduce that $P^{*n} - U = (P - U)^{*n}$.

**Exercise 11.2.** Show that if $P$ is a probability on a finite group $G$ such that $P = P * P$, then there is a subgroup $H$ of $G$ such that $P$ is the uniform distribution on $H$.

**Exercise 11.3.** This exercise is for students familiar with metric spaces. Let $G$ be a group of order $n$.

1. Show that $d(a, b) = \|a - b\|_1$ defines a metric on $L(G)$.
2. Show that $L(G)$ is homeomorphic to $\mathbb{C}^n$.
3. Let $M(G) \subseteq L(G)$ be the subspace of all probability distributions on $G$. Show that $M(G)$ is compact.

**Exercise 11.4.** Show that if $n$ is odd, then the simple random walk on the Cayley graph of $\mathbb{Z}/n\mathbb{Z}$ with respect to the symmetric set $\{\pm[1]\}$ is ergodic.

**Exercise 11.5.** Suppose that one has a deck of $n$ cards.

1. Show that there are exactly $2^n - n$ possible permutations that can be reached after a single shuffle.
2. Prove that a permutation can be obtained after $k$ riffle shuffles if and only if it has at most $2^k$ rising sequences.

**Exercise 11.6.** Show that every eigenvalue of the spectrum of a random walk on a finite group has modulus at most 1.

**Exercise 11.7.** Let $P$ be a probability distribution on a finite group $G$. Show that the following are equivalent:

1. The matrix of $P$ with respect to the basis $\{\delta_g \mid g \in G\}$ is symmetric;
2. $P(g) = P(g^{-1})$ for all $g \in G$.

In this case, the random walk driven by $P$ is said to be *symmetric*. Show that a simple random walk on the Cayley graph of a group with respect to a symmetric set is symmetric.

**Exercise 11.8.** Let $P$ be a probability on a finite group $G$.

(a) Show that if the sequence $(P^{*k})$ converges to $U$, then $P$ is ergodic.
(b) Suppose in addition that $G$ is abelian. Show that the random walk driven by $P$ is ergodic if and only if $|\widehat{P}(\chi)| < 1$ for all $\chi \in \widehat{G}^*$. (Hint: use (a) and the proof of Theorem 11.4.16.)

**Exercise 11.9.** Let $G$ be a finite group. If $f \in L(G)$, define $\widetilde{f} \in L(G)$ by $\widetilde{f}(g) = f(g^{-1})$ for $g \in G$.

1. If $a \in L(G)$, show that $\|a\|_1 = \|\widetilde{a}\|_1$.
2. If $a, b \in L(G)$, show that $\widetilde{a * b} = \widetilde{b} * \widetilde{a}$.
3. If $P$ is a probability on $G$, verify that $\widetilde{P}$ is a probability on $G$.
4. Prove $\|P^{*k} - U\|_{TV} = \|\widetilde{P}^{*k} - U\|_{TV}$ for all $k \geq 0$.
5. Verify that if $P$ is the probability on $S_n$ corresponding to the Gilbert–Shannon–Reeds shuffle, then $\widetilde{P}$ corresponds to the following process. For each position $1, \ldots, n$ of the deck, flip a fair coin. Move to the front all cards in positions for which a heads was obtained, preserving the relative ordering. This is called the *inverse riffle shuffle*.

# References

1. Dave Bayer and Persi Diaconis. Trailing the dovetail shuffle to its lair. *The Annals of Applied Probability*, 2(2):294–313, 1992.
2. Helmut Bender. A group theoretic proof of Burnside's $p^a q^b$-theorem. *Mathematische Zeitschrift*, 126:327–338, 1972.
3. Tullio Ceccherini-Silberstein, Fabio Scarabotti, and Filippo Tolli. *Harmonic analysis on finite groups*, volume 108 of *Cambridge Studies in Advanced Mathematics*. Cambridge University Press, Cambridge, 2008. Representation theory, Gelfand pairs and Markov chains.
4. Charles W. Curtis. *Linear algebra*. Undergraduate Texts in Mathematics. Springer-Verlag, New York, fourth edition, 1993. An introductory approach.
5. Charles W. Curtis and Irving Reiner. *Representation theory of finite groups and associative algebras*. Wiley Classics Library. John Wiley & Sons Inc., New York, 1988. Reprint of the 1962 original, A Wiley-Interscience Publication.
6. Giuliana Davidoff, Peter Sarnak, and Alain Valette. *Elementary number theory, group theory, and Ramanujan graphs*, volume 55 of *London Mathematical Society Student Texts*. Cambridge University Press, Cambridge, 2003.
7. Persi Diaconis. *Group representations in probability and statistics*. Institute of Mathematical Statistics Lecture Notes—Monograph Series, 11. Institute of Mathematical Statistics, Hayward, CA, 1988.
8. Persi Diaconis. A generalization of spectral analysis with application to ranked data. *The Annals of Statistics*, 17(3):949–979, 1989.
9. Persi Diaconis and Mehrdad Shahshahani. Generating a random permutation with random transpositions. *Zeitschrift für Wahrscheinlichkeitstheorie und Verwandte Gebiete*, 57(2): 159–179, 1981.
10. Larry Dornhoff. *Group representation theory. Part A: Ordinary representation theory*. Marcel Dekker Inc., New York, 1971. Pure and Applied Mathematics, 7.
11. John B. Fraleigh. *A first course in abstract algebra*. Addison-Wesley Publishing Co., Reading, Mass.-London-Don Mills, Ont., $7^{th}$ edition, 2002.
12. William Fulton. *Young tableaux*, volume 35 of *London Mathematical Society Student Texts*. Cambridge University Press, Cambridge, 1997. With applications to representation theory and geometry.
13. William Fulton and Joe Harris. *Representation theory*, volume 129 of *Graduate Texts in Mathematics*. Springer-Verlag, New York, 1991. A first course, Readings in Mathematics.
14. David M. Goldschmidt. A group theoretic proof of the $p^a q^b$ theorem for odd primes. *Mathematische Zeitschrift*, 113:373–375, 1970.
15. Marshall Hall, Jr. *The theory of groups*. The Macmillan Co., New York, N.Y., 1959.

B. Steinberg, *Representation Theory of Finite Groups: An Introductory Approach*, Universitext, DOI 10.1007/978-1-4614-0776-8,
© Springer Science+Business Media, LLC 2012

16. Gordon James. *The representation theory of the symmetric groups*, volume 682 of *Lecture Notes in Mathematics*. Springer, Berlin, 1978.
17. Gordon James and Adalbert Kerber. *The representation theory of the symmetric group*, volume 16 of *Encyclopedia of Mathematics and its Applications*. Addison-Wesley Publishing Co., Reading, Mass., 1981. With a foreword by P. M. Cohn, With an introduction by Gilbert de B. Robinson.
18. David A. Levin, Yuval Peres, and Elizabeth L. Wilmer. *Markov chains and mixing times*. American Mathematical Society, Providence, RI, 2009. With a chapter by James G. Propp and David B. Wilson.
19. Bruce E. Sagan. *The symmetric group*, volume 203 of *Graduate Texts in Mathematics*. Springer-Verlag, New York, second edition, 2001. Representations, combinatorial algorithms, and symmetric functions.
20. Jean-Pierre Serre. *Linear representations of finite groups*. Springer-Verlag, New York, 1977. Translated from the second French edition by Leonard L. Scott, Graduate Texts in Mathematics, Vol. 42.
21. Barry Simon. *Representations of finite and compact groups*, volume 10 of *Graduate Studies in Mathematics*. American Mathematical Society, Providence, RI, 1996.
22. Audrey Terras. *Fourier analysis on finite groups and applications*, volume 43 of *London Mathematical Society Student Texts*. Cambridge University Press, Cambridge, 1999.

# Index



B. Steinberg, *Representation Theory of Finite Groups: An Introductory Approach*, Universitext, DOI 10.1007/978-1-4614-0776-8,
© Springer Science+Business Media, LLC 2012